Playing Possum

How Animals Understand Death

Susana Monsó

PRINCETON UNIVERSITY PRESS
PRINCETON *&* OXFORD

First published in Spanish by Plaza y Valdés Editores, 2021, 2022
© Plaza y Valdés Editores, 2021, 2022, © Susana Monsó, 2021

English translation copyright © 2024 by Princeton University Press
Princeton University Press is committed to the protection of copyright and the intellectual property our authors entrust to us. Copyright promotes the progress and integrity of knowledge created by humans. Thank you for supporting free speech and the global exchange of ideas by purchasing an authorized edition of this book. If you wish to reproduce or distribute any part of it in any form, please obtain permission.

Requests for permission to reproduce material from this work should be sent to permissions@press.princeton.edu

Published by Princeton University Press
41 William Street, Princeton, New Jersey 08540
99 Banbury Road, Oxford OX2 6JX

press.princeton.edu

All Rights Reserved

Library of Congress Cataloging-in-Publication Data

Names: Monsó, Susana, 1988– author.
Title: Playing possum : how animals understand death / Susana Monsó.
Other titles: Zarigüeya de Schrödinger. English
Description: Princeton : Princeton University Press, [2024] | Includes bibliographical references and index.
Identifiers: LCCN 2023057783 (print) | LCCN 2023057784 (ebook) | ISBN 9780691260761 (hardback) | ISBN 9780691260853 (ebook)
Subjects: LCSH: Animals—Mortality. | Animal psychology. | Perception in animals. | Animal behavior. | BISAC: SCIENCE / Life Sciences / Zoology / Ethology (Animal Behavior) | SOCIAL SCIENCE / Death & Dying
Classification: LCC QL785 .M58613 2024 (print) | LCC QL785 (ebook) | DDC 591.5—dc23/eng/20240405
LC record available at https://lccn.loc.gov/2023057783
LC ebook record available at https://lccn.loc.gov/2023057784

British Library Cataloging-in-Publication Data is available

Editorial: Rob Tempio and Chloe Coy
Production Editorial: Terri O'Prey
Text Design: Haley Chung
Jacket Design: Haley Chung
Production: Erin Suydam
Publicity: Carmen Jimenez and Matthew Taylor
Copyeditor: Karen Verde

Jacket illustration by Joanna Grochocka / Marlena Agency

This book has been composed in Arno Pro

Printed in the United States of America

10 9 8 7 6 5 4 3 2 1

*For Tote, to whom I owe
so much of who I am*

Death is mundane,
and should we feel the need to theorise about it,
we should do so more mundanely.

—Bob Plant, *The Banality of Death*

Contents

	Foreword, by Mark Rowlands	xi
1	Introduction: The Silence of the Chimps	1
2	The Ant Who Attended Her Own Funeral	8
3	The Whale Who Carried Her Baby Across Half the World	31
4	The Ape Who Played House with Corpses	53
5	The Dog Who Mistook His Human for a Snack	77
6	The Elephant Who Collected Ivory	106
7	The Opossum Who Was Both Dead and Alive	151
8	Conclusion: The Animal Who Brought Flowers to the Dead	206
	Acknowledgments for the Spanish Edition	211
	Acknowledgments for the English Edition	213
	Notes	215
	Image Credits	235
	Index	237

Foreword

Quoth the eponymous *Raven* from Edgar Allen Poe's poem, "Nevermore." To understand that someone has died, one might think, you must understand that *nevermore* will he or she return. *Nevermore* will you see her face again. *Nevermore* will you hear his voice. *Nevermore* will you be together again. Not in this plane of existence anyway.

But *nevermore* is a complex concept. Nevermore amounts to eternity. Never, not as long as time exists, will this dead individual be with us again. Nevermore implies forever, and forever is just another word for eternity. But do we really need to understand eternity in order to understand death? Who among us really understands eternity? Eternity is an infinite period of time—an infinite span of time separates the deceased from her family and friends. And who among us ever really understands the infinite? Perhaps no one. Mathematicians can't even agree if there is any such thing.

Does this mean that no one of us—except perhaps some of the most intellectually gifted humans who have ever lived—understands death? That would be a strange conclusion. We all know what death is. We might not like to think about it. Indeed, some philosophers tell us that, existentially speaking, we flee from it. We refuse to acknowledge it. Try not to think about it. Regard our own deaths as merely distant events in an as yet undetermined future. But, if we do flee our own death in this way, the same philosophers agree, this can only be because, really, and viscerally, we understand death all too well.

We humans are complex creatures. Perhaps we are overcomplicated. We certainly have a tendency to overcomplicate things. Philosophers are the worst of all. In the hands of some, philosophy is the art of overcomplication. There is nothing so simple that some philosophers cannot bend it into shapes of unimaginable, unnecessary, and ultimately implausible complexity. In this, philosophers might be advised to learn intellectual restraint. One way of acquiring this constraint is to learn lessons imparted by individuals who do not share our proclivity for overcomplication. These individuals we know as animals.

In this book, Susana Monsó provides a timely antidote to this unfortunate tendency of philosophers toward overcomplication. Can animals understand death? It is difficult to see how they could not. Any animal that could not understand death would be at a serious disadvantage in the Darwinian struggle for survival. But presumably animals understand eternity, and infinity, no more than we do. So, what is it that animals understand when they understand death? And can this illuminate what we—human animals—understand when we understand death?

This book presents a cogent account of what we—and other animals—understand when we understand death. If we want to understand ourselves, do we look to the philosophers or the animals? Susana Monsó—a philosopher, and a very gifted one indeed—has made a compelling case that we should look to both.

Mark Rowlands

Playing Possum

1

Introduction: The Silence of the Chimps

In November 2009, *National Geographic* published a picture that would capture the imagination of readers and scientists alike. In it one could see Dorothy, a forty-something-year-old chimpanzee, lying on a wheelbarrow that was being pushed by two humans. In the background a group of sixteen chimpanzees huddled behind a fence, each and every one of them staring intently at their fellow (see figure 1). The reason this picture fascinated so many people was that Dorothy was dead, and the rest of her conspecifics, with whom she had cohabitated for her last eight years at the Sanaga-Yong Chimpanzee Rescue Centre in Cameroon, appeared to have gathered to bid her farewell.

Monica Szczupider, the photographer who captured this moment, described it as follows: "Chimps are not silent. They are gregarious, loud, vocal creatures, usually with relatively short attention spans. But they could not take their eyes off Dorothy, and their silence, more than anything, spoke volumes."[1] But what exactly was this silence saying? Is it possible

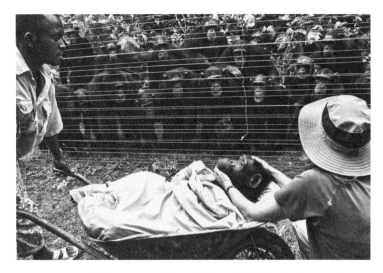

FIGURE 1. "The grieving chimps," photo by Monica Szczupider.

that the chimps were experiencing something similar to our grief over the loss of a loved one? Could they understand what had happened to Dorothy? Did they perhaps know that that very thing would happen to them sooner or later?

This picture sparked such interest that it led various scientists to publish similar cases that they had witnessed throughout the years but had not yet documented, and many others began to pay closer attention to the behaviors surrounding death of the animals they were studying. With this, a new discipline was born: *comparative thanatology*, which aims to study how animals react to individuals who are dead or close to dying, the physiological processes that underlie their reactions, and what these behaviors tell us about the minds of animals. Although the focus was originally on primates, the last few years have seen an explosion of publications on the topic. As a result, there is an increasing number of articles on the thanatology of species that are far removed from monkeys and apes, such as elephants, whales, horses, crows, and even insects.

This interest in how animals relate to death is part of a growing scientific trend that addresses the extent to which animals possess capacities traditionally believed to be solely human. A mounting number of studies suggest that many animal species are endowed with at least rudimentary forms of old guarantors of human uniqueness, such as numerical cognition, rationality, morality, language, or culture.[2] The idea of human beings as an entirely separate, more-than-animal species is becoming less and less tenable by the day. Naturally, the question of whether animals possess a notion of mortality becomes relevant in this context, for throughout the ages humans have thought of themselves as the only species blessed—or cursed—with an understanding of death.

Comparative thanatology—the study of animals' relation to death—is a discipline located at the intersection of *ethology* and *comparative psychology*. Ethology is the branch of biology that focuses on the study of animal behavior, and it shares with comparative thanatology a predilection for field studies carried out in more or less natural settings. Comparative psychology, in turn, aims to study animal minds experimentally and compares how different species deal with similar problems and what cognitive mechanisms they use to resolve them. Comparative thanatology shares with this discipline an interest in animal psychology, and also makes use of many of its studies to inform the debate on how animals experience and understand mortality.

This book is not written by an ethologist nor by a psychologist, but by a philosopher. This may surprise you, if your image of the philosopher corresponds to a bearded old man who smokes a pipe and sits in his armchair reflecting upon the meaning of life. I won't deny that this description fits some of us, but the truth is that philosophy is a very heterogeneous discipline, and not only are there philosophers from a wide variety of age groups, genders, and ethnicities, but many of us also spend our time studying topics—such as climate change, terrorism, video games, medicine, or porn—that don't fit the popular image of what philosophers like to ponder.

The variety of topics that philosophers study reflects certain peculiarities that this discipline has and that distinguish it from others. In contrast to other branches of science and the humanities, philosophy lacks a predetermined object of study. There can be philosophy of anything because philosophy is a method, a way of looking at the world and reflecting on it, rather than the study of a particular, concrete phenomenon. This allows philosophers to be in a constant dialog with other areas of knowledge, to move with ease from one discipline to another, to take nothing for granted, to question every assumption, and to offer refreshing and innovative points of view that can serve as catalysts for any debate.

This book is framed within a relatively young branch of philosophy known as *philosophy of animal minds*. Although philosophy of mind goes back, at the very least, to ancient Greece, throughout history it has focused almost exclusively on the human mind. Philosophy of animal minds vindicates the study of the minds of animals, not just to understand ourselves better, but also as an end in itself, given the assumption that the psychology of other species is interesting independently of what it can teach us about our own. In turn, this discipline works in dialog with science, reflecting on the methodologies with which we study the behavior and cognition of other species, identifying potential biases, and aiming to provide conceptual clarity.

Comparative thanatology, as a discipline that has existed for barely a decade, is very much in need of a philosophical outlook that can help to identify the hidden assumptions that may be biasing its research, as well as to clarify the meaning of its key concepts. This book centers specifically on identifying and removing the anthropocentric biases that underlie the investigation of how animals relate to mortality. Moreover, the key concept on which I focus, and which makes up the backbone of the overarching argument, is the concept of death. What exactly does it mean to *understand* death? Is the concept of death something binary, an all-or-nothing matter, or can we conceive it as a spectrum, as something that admits higher or lower degrees of complexity? Would it make sense to talk about different concepts of death that capture the perspectives of different species?

An important part of the work that I carry out in this book is, therefore, one of conceptual analysis. However, this does not merely consist of clarifying the language being used, for through such an analysis one can arrive at conclusions about the world. For example, in order to determine whether the experiments that demonstrate altruistic behaviors in animals are evidence that animals are moral, we need to start from a clear characterization of what it means to be moral. The same applies in this case. Through an analysis of what it means to have a concept of death, we can look at existing evidence from a different perspective. What's more, this analysis will allow us to clearly delineate the *cognitive requirements* for understanding death; the psychological architecture that an animal must be endowed with in order to have an awareness of mortality. Knowing this, we can then look beyond comparative thanatology and consider what other fields, such as evolutionary biology, can tell us about the extent to which this capacity is likely to be found in nature.

Do animals understand death? In this book I use the conceptual and argumentative tools that philosophy makes available to us in order to analyze the empirical evidence that has been accumulating in the field of comparative thanatology during the past decade, and thus provide an answer to this question. As we shall see, since its birth this discipline has been characterized by certain anthropocentric biases that have led thanatologists to intellectualize the concept of death and place an excessive emphasis on grief as an emotional response to others' demise. Locating and eliminating these biases will allow us to see that the concept of death requires little cognitive complexity and that there are multiple ways in which animals can emotionally react to death and learn about it. If my arguments in this book are correct, the concept of death is much easier to acquire than has usually been presupposed and is likely to be widespread in the animal kingdom.

Perhaps all of this sounds outlandish to you, if you are not used to hearing about the concepts or emotions of animals. If this is so, I would ask you to put some trust in me, for this book is written for readers with no previous knowledge of the matter and does not require you to master any notion regarding animal psychology. On the other hand, perhaps you belong to the group of people who doubt that animals even have minds. If that is the case, today's your lucky day, for here I will not only discuss animals' relation to death, but I will also tackle both philosophical arguments and empirical evidence that support the notion that humans are far from the only animals with a mental life. Therefore, if you are a skeptical reader, you should find in this book, at a minimum, some food for thought.

In what follows, we will begin with philosophy and delve deeper and deeper into comparative thanatology and its related empirical sciences. I have aimed to keep technical distinctions to a minimum and, when they were absolutely necessary, I have attempted to explain them with care, along with a touch of humor wherever possible (though hilarity is regrettably not always guaranteed). To those readers who struggle a bit with philosophy, I ask for patience. For those who came looking for stories about animals, I promise they will arrive. And without further ado, dear reader, I would like to thank you for choosing this book. I sincerely hope that you will enjoy it and learn something about how animals understand death—and maybe a bit about how we do as well.

2

The Ant Who Attended Her Own Funeral

When I was eight years old, Santa Claus brought me a microscope. I had asked for it after seeing a TV ad in which some kids dressed in lab coats and goggles examined all kinds of common objects that became fascinating upon being magnified. It was a present that I was extremely excited about—at least in theory.

When I began to play with my new acquisition, I soon realized that it was much more boring than the ad had led me to believe. The microscope included some samples for examination, but they were few and far from the level of awesomeness I had been promised. And the objects we had lying around the house were too opaque, too big, or too dull to merit investigation under my microscope. I needed something with more allure.

That is when I came up with the idea of examining an ant up close. Who hasn't felt astonished by the alien face of these insects, their massive eyes, their ferocious mandibles, their unsettling antennae? That promised to be the thrilling experience I was looking for.

My plan just had a tiny hole: I needed the ant to be dead. Otherwise, she* would be wriggling nonstop and it would be impossible to get a proper look at her. And the only method I knew of to kill ants—the classic stamping—would have worked perfectly if my intention were to examine puréed arthropod, but I was more interested in the intact subject.

So, one morning I got one of the little sample tubes that came with my microscope and I trapped an ant inside it with the idea of leaving her in there until she ran out of oxygen. I'm ashamed to admit it, but I spent all day with the little tube inside my pocket, taking it out from time to time to check whether the ant had decided to contribute to the advancement of science and been kind enough to kick the bucket. But no, there she was. Alive. Very much alive. Neither she nor I knew it in that moment, but that inconvenient persistence in surviving was what would ultimately save her life.

Later that day, with the Colacao† remains all dried up on my moustache, I became fully conscious of what I was doing. For some reason, I put myself in the ant's shoes, and was overcome by unbearable sadness (I was a curious kid, but not completely heartless). How could I leave her to die? Who was I to take away this innocent being's life? Suddenly it had become an act of unacceptable cruelty.

I remember going up to a beautiful rosebush in our garden and carefully opening the tube to let the ant run free, at last, on one of its leaves. The poor thing got out of there slowly and

---

* Throughout the book, as is customary in contemporary analytic philosophy, I use "she" as a gender-neutral pronoun whenever the sex of the animal I'm referring to is unknown.

† A Spanish alternative to Nesquik, and a far superior one, despite what the haters might say.

seemingly disoriented, as though asking herself why she had been abducted in such a bizarre way. It's more than likely that she died shortly thereafter, but in that moment, I felt like I had saved her. Like I had done the right thing. Like a bona fide Little Miss Captain Marvel.

That moment of connection and empathy with the ant, which has stayed in my memory with such vividness, likely contributed to shaping the person I am today. Although at that time I wanted to be a painter, I would end up pursuing a career in animal ethics.

The arguments that follow will have ethics as their constant backdrop, but this is not the reason I am telling this story. In this book, I'm going to talk about how animals experience and understand death. Naturally, we humans are animals too, and as such, our own death concept is going to be very present throughout. In fact, it will serve as a counterpoint and a guide in my reflections on what understanding mortality amounts to.

My anecdote with the ant is relevant because of what it reflects about the concept of death that I had back then. I knew that the ant belonged to the class of things that can die (in contrast to, for instance, a stone, which is not alive but which also doesn't die). I could predict that once the ant died she would stop moving, which would allow me to examine her under the microscope. I also had some knowledge of what could cause the death of an ant (although, judging by the highly inefficient method I used, one must admit that it wasn't a very sophisticated knowledge). Moreover, I was capable of determining that the ant was still alive, and, had the ant died, I would have recognized her death. Last, I was capable of empathizing with her and seeing the death of the ant as a tragedy, as something negative that had to be prevented. All of this indicates a fairly complex concept of death.

Ants and Death

We can see some similarities between my story and how ants themselves would react to the death of a fellow. First, ants are also willing to mobilize themselves in order to prevent one of their sisters from dying. In a recent experiment,[1] researchers used nylon snares to trap an ant under some sand, simulating a situation that happens often in nature when these insects are immobilized by collapsing debris.* The scientists found that the other ants would attempt to free the trapped one. At first, they would dig in the sand and pull on the ant's limbs to try to get her out of there. When this strategy didn't work, they would switch to focusing on the snare, biting on the nylon thread until they broke it and thus saving their mate.

Second, ants are also capable of discriminating when a conspecific has died. When an ant perishes inside the colony, the rest identify it and will diligently proceed to extract the corpse from the nest, a behavior known as *necrophoresis*.

Despite these similarities, we can state with a reasonable degree of confidence that ants lack a concept of death. The ant who rescues her trapped fellow is not anticipating her death and trying to avoid it—she is responding to a chemical call for help emitted by the ant in peril. In fact, if we anaesthetized the trapped ant, the others would not rescue her. They need this chemical call in order to initiate their helping behavior.

The case of necrophoresis is similar. It's not that the ants who engage in it have understood that the other ant is dead—rather,

---

* Throughout the book I will occasionally refer to experiments that, like this one, can be criticized from a moral perspective. In order not to distract from the main argument, I will not comment on this further, but I do want to note that my reference to any particular study does not imply an ethical endorsement of its methods.

they are offering a pre-programmed response to a chemical stimulus. Necrophoresis is initiated upon the detection of the oleic acid that is given off by these insects' carcasses. When other ants detect it, they immediately respond to the dead ant as "garbage that must be taken out of the nest." Due to the rigidity of this mechanism, it would be easy to trick them. If we were to grab a live ant and put a drop of oleic acid on her abdomen, the others would treat her as though she were dead. That is, they would pick her up and take her out of the nest.[2] They would not be capable of inferring that they're dealing with a live individual even if the "corpse" were incessantly moving about and trying to shoo off her captors.

The fact that these behaviors are driven by the detection of chemical components doesn't mean that they aren't somewhat flexible. The ants who rescue their trapped conspecific show a certain understanding of the situation: if a strategy doesn't work, they are capable of switching tactics and they end up processing that the problem is the snare, despite this not being an element that they would find in nature.

Something similar happens with necrophoresis. Even though it's a behavior that has at its roots a stimulus-response mechanism, it's not as simple as it is often characterized.[3] In fact, oleic acid is not the only stimulus that can trigger necrophoresis. Other chemicals produced during putrefaction can also serve that function. Some ant species can even react before the decomposition starts, meaning that the absence of a chemical signal of life (for example, certain pheromones) also appears to be sufficient. There are species that distinguish between fresh and old corpses, and while they might bury the former, the latter are extracted from the nest.[4] Context also appears to be important: a decomposing ant encountered during foraging does not trigger the necrophoresis process, unless she is located very close to the nest. And if a carcass cannot be extracted from the colony due to some environmental factor, like a frost or a flood, the ants can opt for other elimination methods, such as burying or even cannibalism.[5]

Faced with the complexity of the ants' behavior, how can we be so sure that they lack a concept of death? In order to answer this question, we need to have a clear idea of what concepts are.[6] We can think of them as the basic components of many of our thoughts. For instance, the thought "The table wobbles" is made up of the concepts of *table* and *to wobble*. Although, as we will see, not all forms of thinking require concepts, a creature that wasn't at all capable of thinking could never possess concepts. Therefore, before we go into greater detail on what it would mean for animals to have concepts, we need to take a short philosophical detour to face a more general question: do animals have minds?

The Philosophical Dispute over Animal Minds

In 1982, philosopher Donald Davidson published a famous essay in which he defended a negative answer to the question of whether animals have a mental life.[7] Davidson asks us to imagine a dog who is chasing the neighbor's cat through his garden. The cat runs straight at an oak tree, but in the last moment he swerves and goes up a different tree, in this case a maple. The dog misses this last maneuver and obstinately runs toward the oak tree. He stops at its base, rests his front paws on its trunk, and begins to bark at its top. If we were to witness such a scene, we couldn't help but proclaim: "The dog believes the cat is in the oak tree!"[8] That is, we would explain the dog's behavior by attributing a *belief*—a type of thought—to him. Davidson, however, identifies two big problems with this kind of attribution.

On the one hand, there is a problem that has to do with the limits of our knowledge. It's impossible for us to adopt the perspective of the dog, which prevents us from directly accessing his way of understanding the world. But having access to the way an individual understands the world is fundamental when it comes to attributing beliefs to her. For example, we can attribute to Lois Lane the belief that she works with Clark Kent but not the belief that she works with Superman, despite the fact that Clark Kent and Superman are the same person. This is due to the fact that Lois Lane *doesn't know* that Clark Kent and Superman are the same person, which is why we ought to use the name "Clark Kent" when we discuss how she perceives her colleague.

Now, the problem with the case at hand is that we lack any way of knowing what exact form the dog's belief would take. We can't know, for instance, if the dog believes that the cat went up "the oldest tree in the garden," "the tree that smells the best," or "the tree that the cat went up last time." The attribution of concrete beliefs to the dog would encounter this insurmountable problem.

On the other hand, for Davidson, beliefs are never held in isolation, but rather all beliefs need other beliefs in order to have meaning. A dog can't believe of an object that it is a tree without a series of general beliefs about trees: that they grow; that they have leaves, branches, and roots; that they need light and water, etc. And the same goes for the cat: believing that this is a cat implies believing that she's a mammal, that she has four legs, that she meows, etc. Even though there is no fixed list of general beliefs that one has to have in order to have the belief that this object is a tree or a cat, without the attribution of at least some of these general beliefs, Davidson tells us, it stops making sense to attribute to the dog a belief that the cat is in the tree. However, once again, we lack a way of knowing which of these beliefs the dog has, if indeed he has any of them.

Few philosophers have been convinced by Davidson's argument.[9] It has been pointed out, first, that the problems with the attribution of concrete beliefs to animals also come up, although less saliently, when we attribute beliefs to other humans. We can never know exactly what is going on in the minds of others, nor can we access beforehand and with all manner of details how they conceive the world. And yet, this does not prevent us from making approximate attributions based, for instance, on what we know about their personality or their life history. Perhaps Lois Lane thinks of her colleague as "the dork with the glasses" and not as "Clark Kent," but it still seems appropriate to use the latter term instead of "Superman," knowing as we do that she still has no inkling of what's up with him.

At the same time, it has been argued that Davidson is excessively pessimistic when it comes to describing what we can know about animals. If we study their behavior under very controlled conditions, we can come to know quite a lot about how they understand the world, which in turn allows us to delineate their beliefs much more than Davidson acknowledges.

It has also been argued that animals' beliefs don't have to be made up of concepts, as Davidson assumes. It could be the case, for example, that the dog thinks of the cat not as a four-legged meowing mammal, but in terms of the actions he allows: for instance, as a chase-able or eat-able thing.

Last, it has been emphasized that, even if we were to grant Davidson's argument, all it ultimately allows us to conclude is that we don't know *what* beliefs are held by animals, but it doesn't give us any reason to think that they have no beliefs at all. The problem of exactly what is going on in animals' heads is a different problem from the problem of whether they have a mind at all.

However, for Davidson, it's not merely that we lack a way of knowing *what* beliefs animals have, but also that the very idea that animals have a mind is problematic. Here is the argument he offers. An individual who has beliefs has to be capable of being surprised, for surprise consists of noticing that reality is not like one *believed* it was. For instance, if I'm surprised when I see that my bank account is in the red, this is because I can process the contrast between the money I *believed* I had and what I actually have. For an individual to have the capacity to be surprised, and thus for her to have beliefs, Davidson argues that she has to have the *concept of belief*—she has to be capable of understanding that one thing is what one believes and a very different one is the way the world actually is. By being surprised at the state of my bank account, I show that I understand the difference between the fantasy I built in my head and the harsh reality of my circumstances.

Now, for Davidson, language is necessary in order to form the concept of belief. Language allows us to contrast what we believe with what others believe, thus building the notion of an objective reality that is independent of all our subjective beliefs. Given that only humans have language—or at least one that is sufficiently sophisticated for this task—only humans can have the concept of belief, and thus beliefs. And because, for Davidson, beliefs are basic and fundamental for all kinds of thought, only humans have minds.

Once again, few philosophers have been convinced by this argument. Davidson has been accused, for example, of putting the cart before the horse. Although there is a connection between the capacity to have beliefs, the concept of belief, and language, the relationship between the three is more likely to be the inverse. That is, having beliefs is probably necessary to develop a language and the concept of belief, and not the other way around. Similarly, assuming that beliefs require both language and a concept of belief implies that prelinguistic children lack beliefs. This, in turn, makes it very difficult to explain how they can acquire a language. Without beliefs, with a head that is completely devoid of thoughts, it becomes very complicated to see how one could come to acquire the names of things and learn to express ideas.

Therefore, Davidson's arguments are not too convincing. But so far I have shown only that this specific philosopher was wrong. What positive arguments can we offer in favor of the existence of animal minds? It's common to distinguish three main strategies: the argument from analogy, the inference to the best explanation argument, and the argument from evolutionary parsimony.[10]

The *argument from analogy* focuses on a certain property X that we know to be linked to mental abilities in humans and aims to conclude, by way of an analogy, that animals who show that property X also have a mind. For instance, problem-solving and behavioral flexibility are properties that, in humans, depend on our mental capacities. We can conclude, by analogy, that those animals who are capable of solving problems or who show flexible behaviors also have a mind. In fact, there is a huge and growing quantity of studies in comparative psychology that show that many nonhuman species are capable of making inferences, remembering the past, anticipating the future, planning, innovating, adapting to changing circumstances, and many other behaviors that in us require some form of thinking.[11]

The *inference to the best explanation argument* is based on the idea that assuming that animals have a mind is the best way to explain certain behaviors that we see in them, such as those behaviors that are apparently misguided or do not fit reality. Given that the cat is not actually in the oak tree but in the maple, the best explanation of the behavior we see in the dog is that he *thinks* the cat is in the oak tree. The inference to the best explanation argument also makes use of the idea that not only can we *explain* certain behaviors better by attributing mental states, but that we can also make *predictions* based on these mental states. For example, if we assume that the dog has the *desire* to catch the cat and the *belief* that the cat is in the oak tree, we can predict that, when he realizes that the cat is actually in the maple, he will first act surprised and will then run toward where he has now learned that the cat is. If our predictions are confirmed, the best explanation of this success will once again imply that animals do indeed have minds.

Last, the *argument from evolutionary parsimony* is based on our knowledge of biological evolution and natural selection. All species are related with one another. We can pick any two species, regardless of how different they are, and they will always have a common ancestor. The closer their kinship, the more the two species will resemble each other. This also applies to mental faculties. Those species that are close to each other in the tree of life will probably share many of those faculties. When we see, in species that are closely related to our own, behaviors that we know to involve our mental capacities, we can assume that those species will also have a mind—this is the most parsimonious explanation. This argument implies that, the higher the degree of kinship of another species with our own—or with species whose psychology has been satisfactorily proven—the higher the chances that they will have a mind.

Animal Concepts

Let's assume then that at least some animals are capable of thinking despite lacking a language. This doesn't necessarily mean that they possess concepts, for some forms of thought may be nonconceptual. We can imagine, for instance, a squirrel who is planning how to get from the branch she's currently standing on to a branch from the tree in front. To do this, in principle she doesn't need a concept of branch nor a concept of tree. It might be enough for her to have, for example, the capacity to think in images; to make a mental map of the tree where she can imagine and try out different routes. This doesn't imply that squirrels lack concepts, simply that they don't need them for this concrete form of thinking. For us to be able to say that an animal has concepts, we have to show not just that she's capable of thinking, but also that she has certain specific abilities.

First, an animal with a concept will have the capacity to discriminate the entities to which that concept applies with a certain degree of reliability. For instance, if Carla has the concept of dog, Carla should be capable of distinguishing dogs from other entities, such as shoes or umbrellas, and from other living beings, such as frogs or bats. Carla might not be completely infallible when it comes to discriminating dogs. If the only dogs she's seen in her life are Pyrenean mastiffs, the first time she comes across a chihuahua she might think she has encountered some other animal, but this will not indicate that she lacks a concept of dog. Having a concept, thus, does not imply being free of the possibility of making a mistake in a classification, but it should come with the capacity to learn and improve one's concepts. If we were to explain to Carla that chihuahuas are also dogs, she should be capable of extracting them from the mental box of "possible rodents" and include them in the one of "dogs," such that the next time she sees one she will be capable of classifying her correctly.

Second, possessing a concept doesn't only mean the capacity to discriminate the entities to which that concept applies, but also entails a certain understanding of what philosophers call the "semantic content" of that concept, that is, its meaning. Continuing with Carla, she will not only be capable of distinguishing dogs from other things, but will also have some knowledge regarding what it means to be a dog. She may know, for example, that dogs are mammals, that they have four legs, that they bark, that they are covered in fur, or that they would lose their head over chasing a ball. This, as Davidson already anticipated, means that all concepts are connected with other concepts in a kind of semantic net. The concept of dog is connected with concepts such as that of *mammal* or that of *barking*.

At the same time, the content of each concept is not static or permanent. Rather, it will vary depending on the moment in time, the culture, and the person we are dealing with. Thus, the concept of dog that we have nowadays is very different from the one we had two thousand years ago, when scientific knowledge was much less advanced. And the concept of dog that we have in Europe, where dogs are fundamentally companion animals, will differ from this concept in other societies, where they might for instance be seen as a culinary delicacy. Finally, my current concept of dog—after having lived with one for sixteen years—will be different from the one I had when I was a kid and found them terrifying. Likewise, my concept will be different from my cousin Almudena's, given that she's a veterinarian and is thus familiar with many facts about the anatomy and physiology of dogs that are completely unknown to me.

Third, having an understanding of the semantic content of a concept also allows one to make inferences with that concept. For instance, imagine that we want to trick Carla into believing that our cat Winifred is a fox terrier, for which we dress her up and give her the doggiest appearance we can. Although Carla may initially be fooled—if we created a brilliant disguise and Carla was not the sharpest tool in the box—were her concept to be robust, she would soon realize that the ways in which Winifred moves and wags her tail are not consistent with what's expected of a dog. The moment Winifred meowed or unsheathed her sharp claws, the spell would definitively be broken and Carla would infer that Winifred is actually not a dog. And also perhaps that we're a bit cruel for having so blatantly tried to fool her.

Fourth, concepts are not linked to a specific sensory stimulus. Carla's concept of dog doesn't just apply to dogs of a specific breed or this or that fur color—she can recognize as dogs animals with very different appearances. Her concept also isn't necessarily linked to visual stimuli—she could recognize a dog using her other senses too. For instance, if she were to hear barking or whimpering, she would know that there's a dog nearby. If she entered a house and it smelled of dog, she would deduce that the owners have a pet. If in the middle of the night she felt a wet nose pressing against her arm, she would know that her dog has come to try to wake her up. In fact, Carla doesn't even need a canine sensory stimulus in order to think about dogs. She is capable of having thoughts such as "I would like to adopt another dog" or "I wonder what my grandma's dog is doing right now" without having a dog in front of her. Her concept is, therefore, fairly independent of sensory stimuli.

Last, concepts do not generate a fixed behavioral response. The manner in which Carla reacts to a dog will depend on the complexity of her concept, her mood, her personality, her tastes, her life history, and whatever beliefs and desires she has at that moment. Perhaps one day when she encounters a dog in the park she will walk up to her and pet her because she knows from experience that dogs tend to be open to those kinds of interactions. But if the canine reacts by baring her teeth, next time Carla might think twice before approaching one or will try to do it in the least menacing way possible. Carla's reactions, in turn, will contrast with those of other people who also have the concept of dog. Maybe Lidia, who as a child had a traumatic experience with a German shepherd, would never dare to approach one, and Bea, who has a terrible allergy to dog fur, would also avoid them at all costs.

Concepts, therefore, have six main characteristics: (1) they allow us to discriminate the entities to which they apply with some reliability, (2) they entail a certain amount of knowledge regarding the semantic content of the term in question, (3) they vary across individuals, cultures, and moments in time, (4) they allow us to make inferences, (5) they are not linked to concrete sensory stimuli, and (6) they don't generate a fixed behavioral response. If an animal doesn't display these characteristics in her abilities and behavior, it means she lacks concepts.

Stereotypical and Cognitive Reactions to Death

Having seen what it takes to possess concepts, we can finally return to the example of the ants and discover why we can be fairly sure that these insects are not operating with a concept of death. This has to do with the fact that their behavior evidences *stereotypical* rather than *cognitive* reactions to this phenomenon. To explain the difference between stereotypical and cognitive reactions to death, I shall focus on the example of necrophoresis—ants' tendency to remove their dead fellows from the nest—for it offers the most direct evidence of a possible concept of death in these insects.

As I have pointed out, for ants to have a concept of death, they would have to be capable of discriminating the entities that fall under this concept with some reliability. That is, they would have to be fairly good at distinguishing dead from live ants. We have already seen that this is something they can do thanks to the presence or absence of chemical markers of life and death, such as oleic acid. However, their discrimination of dead ants is completely dependent on these chemical markers. That is why it's so easy to fool them and make them treat a live ant as though she were dead. As far as we know, they are not capable of realizing when they have made a mistake in their discrimination. The stunned ant attending her own "funeral" has no way of letting the others know that they've made a mistake, that she's actually not dead.

Even though the ants' behavior is not as simple as it has sometimes been characterized, we lack any indication that these insects understand *what it means to be dead*. These animals come with a set of "instructions" that are activated upon the detection of sensory stimuli linked to death and other environmental circumstances. But they are not capable of inferring that the ant is alive when they obtain contradictory signs, nor are they capable of learning to improve their discrimination of death. There are also no reasons to believe that they can have thoughts about dead ants, let alone thoughts in the absence of a concrete sensory stimulus.

The necrophoric behavior of the ants, together with the anecdote with which I began this chapter, can be used to distinguish two classes of reactions to death that we can see in the animal kingdom: stereotypical ones and cognitive ones (see table 1).

TABLE 1. Main differences between stereotypical and cognitive reactions to death

Stereotypical reactions to death	Cognitive reactions to death
Innate	Learned
Rigid	Flexible
Automatic	Under cognitive control
Linked to concrete sensory stimuli	Not linked to concrete sensory stimuli
Same individual, same reaction	Same individual, different reactions
Different individuals, same reaction	Different individuals, different reactions
Clearly adaptive	Not necessarily adaptive

Ants represent the class of *stereotypical* reactions: those behaviors that are automatically activated upon the sensory detection of a death-related stimulus. They are genetically inherited behaviors that don't require any learning, for the animal is born with a predisposition toward them. The most common stereotypical reactions to death are *necrophoresis* (which, as we have seen, consists of transporting individuals that smell dead out of the colony) and *necrophobia* (which consists of feeling aversion toward corpses and physically avoiding them). Both reactions are usually triggered upon the sensory detection of death markers and allow organisms to reduce the risk of illnesses associated with the presence of a deceased individual in their surroundings. They therefore have a clear function that usually pertains to the prevention of infections.

If stereotypical behaviors have evolved with such rigidity, it is because, in the environment where the species lives, this protective function gives them a clear evolutionary advantage. Consequently, the fact that they imply little cognition doesn't usually pose a problem. Ants' habitats don't commonly include malicious humans with the intention of spraying oleic acid on live ants. In the ants' natural environment, oleic acid is a very reliable marker of death. This trait allows this adaptation to fulfill its function without the ants needing to be capable of reasoning about death, but it has the disadvantage that, on those rare occasions in which the environment does include a malicious human with lots of time on her hands, these insects are not capable of inferring that the ant who smells dead but is moving is not actually dead.

The anecdote of the little girl who wanted to look at an ant with her microscope represents the second class of reactions to death: the ones we can call *cognitive*. In contrast to the stereotypical ones, cognitive reactions are not genetically inherited, but instead depend on learning and on the life history of each individual. Human children are not born with a predisposition to put ants into sample tubes to observe them later under a microscope. Although my behavior was influenced by my genes, it was also the result of the environment in which I grew up and my life experiences. Had I grown up in some other environment, such an outlandish idea might never have occurred to me and I might have spent that time on some other of my preferred activities, like playing video games or pestering my younger brother.

Cognitive reactions are also not linked to concrete sensory stimuli. In fact, there doesn't have to be a dead individual in the surroundings in order for someone to have a cognitive reaction to death. In my story, the ant didn't end up dying, so there was no corpse triggering my behavior. However, I was able to think ahead about the ant's death and reason, first, about how to bring it about, and, later, about how to prevent it.

Cognitive reactions to death have not been directly shaped by natural selection, so they lack a clear function and don't necessarily come with an evolutionary advantage. However, they may be the indirect result of adaptations that do have an important function. My decision to free the ant, for instance, emerged from a tendency toward empathy that is innate in human beings and that is believed to have the function of facilitating parental caregiving, as well as cooperation and group living.[12] Although not all humans extend their empathy to ants, the majority of us are born with a capacity for empathy.

Given that they are not linked to concrete stimuli and that they lack a strong genetic base, cognitive reactions to death are also characterized by high interindividual variability. The Susana from back then would have seen a dead ant as an opportunity to use her new microscope, whereas someone else might have seen the corpse as something to sweep away, or as something unworthy of attention. Even in one and the same individual we can see very different cognitive reactions, depending on the moment and the situation. What the morning Susana thought was a brilliant idea horrified the afternoon Susana—even though she was still the same kid!

Despite the importance of the distinction between cognitive and stereotypical reactions, it's crucial that we don't interpret it as a strict dichotomy. In some animals, such as ants, we see that stereotypical reactions to death are dominant, but in other animals we can see a mix of both. For instance, a rat that comes across the decomposing carcass of a fellow rat will proceed to bury her. This is a stereotypical reaction triggered by the olfactory detection of cadaverine and putrescine, which, as can be guessed from their names, are two chemicals that emerge from the decomposing process. We know that this behavior depends on the olfactory detection of these death markers because rats are also prone to burying pieces of wood sprayed with cadaverine and putrescine, and rats who lack a sense of smell do not carry out this behavior.[13] However, this behavior is not as rigid as the one we see in the ants. For a rat to be inclined to bury a live conspecific, it's not enough to spray the latter with these chemicals—the victim must in addition be anaesthetized. This suggests that the reaction triggered by the smell of death is not sufficiently strong to compensate clear signs that a rat is alive and well.[14]

It's likely that there are stereotypical reactions to death in a great many species, at least to a certain degree. In fact, we know that necrophobia is widespread in nature, due to decomposing corpses being a paradise for pathogens.[15] For this reason, a putrefying carcass is something that for many animals smells bad, appears disgusting, and generates an inclination to stay as far away from it as possible.

Necrophobia is also present in our species. I invite intrepid readers to google "putrefaction" and try not to feel an immediate aversion to the images that appear on the screen. The disgust we feel is a very important mechanism that serves to protect us from possible diseases, and it's so reliable that a mere picture is sufficient to trigger it, even if what is generating it does not pose a real threat in that moment. In this case, it's a reaction triggered by a visual stimulus, but humans also have stereotypical reactions associated with olfactory markers of death. In a series of recent experiments, the smell of putrescine was found to not only seem repugnant to us, but also to reduce our reaction times, make us walk faster, and increase the probability of having both thoughts related to feelings of danger or threat and feelings of hostility toward people who don't belong to our social group.[16] The smell of putrescine thus works as an alarm signal and automatically activates our fight or flight mechanisms. This is due not only to the fact that corpses are a source of pathogens, but also to the fact that a carcass in one's surroundings is an indication of a potential threat to our physical integrity.

If stereotypical reactions like these can occur in us, the queens and kings of cognitive reactions to death, it's clear that it can't be a strict dichotomy. It's important to remember this, because the fact that an animal shows some stereotypical reaction to a corpse doesn't necessarily mean that she's incapable of developing a concept of death. That being said, cognitive reactions will be our focus of attention, for they are a possible indicator of a capacity to learn about mortality, and thus of developing a concept of death. The species we have to look at, therefore, are those that show high variability and flexibility in their interactions with animals who are at risk of dying or already dead, which would be a sign that their behaviors are not triggered by rigid stimulus-response mechanisms such as those that underlie stereotypical reactions.

In recent decades, scientists have documented the reactions of many nonhuman animals to their conspecifics' deaths, and we have evidence of this variability and flexibility in various species. In one of the most famous examples, a number of chimpanzees who live in a sanctuary were observed reacting to the corpse of an adolescent member of the group.[17] The reactions were very diverse (see figure 2). Some chimpanzees inspected the carcass with apparent interest. Others simply sat around the body and observed it in silence. Some gave alarm calls. Others reacted aggressively, engaging in dominance displays and even attacking the remains. After a while, the human caretakers called the chimpanzees to the feeding building and most of them left the place where the body lay, with the exception of two females, mother and daughter, who remained next to the corpse. The mother, who had had an especially close relationship with the deceased, carefully examined his face and then

FIGURE 2. Chimpanzees reacting in various ways to the corpse of a member of their group.

grabbed a grass stem and began to clean his teeth. She spent several minutes engaged in this task, all the while remaining under her daughter's attentive gaze.

The chimpanzees in this example were clearly showing cognitive reactions to death, that is, reactions that were mediated by cognitive and emotional mechanisms, as well as by their personalities or past experiences. This is why we see such different behaviors despite them being members of the same species. But a cognitive reaction is not the same as a concept of death. The chimps might have interpreted the corpse as being asleep, or perhaps as a curiosity of sorts. The following questions thus emerge: did the chimpanzees know what had happened to their fellow? Do chimps have a concept of death? My aim is to convince you that the answer to these questions is likely to be positive. In fact, as I will argue, we don't have to restrict ourselves to animals as cognitively complex as chimpanzees in order to find a concept of death in nature.

Thus far, I have established two fundamental bases for the argument I will construct in the following chapters. First, I have delineated the distinction between stereotypical and cognitive behaviors. Although many animals, such as ants, show stereotypical reactions to death, my attention in the following chapters will be focused on cognitive reactions. More specifically, I will tackle the issue of whether animals are capable of a specific kind of cognitive reaction to death: reactions mediated by a concept of death. Second, I have defined what concepts are, which was crucial, given that I'm interested in the *concept* of death. In what lies ahead, I will spell out what an animal must understand not only for her concepts to count as such, but for her to also have a concept *of death*. But now it's necessary to identify some of the biases that might obstruct this task and cloud our reasoning.

3

The Whale Who Carried Her Baby Across Half the World

In the summer of 2018, a female orca nicknamed Tahlequah appeared in news outlets from all over the world as the protagonist of an apparent tragedy. After a seventeen-month-long pregnancy, Tahlequah had given birth to a calf who had managed to live for only half an hour. This was sad news for Tahlequah's pod, who are part of an endangered population and hadn't seen a live birth for three years. The mother appeared incapable of accepting it. She was seen carrying the baby's body while she swam, balancing it on her snout to prevent it from sinking. This behavior, which could patently be interpreted as a display of grief, lasted an astounding seventeen days, during which Tahlequah followed her family across more than a thousand miles, her dead offspring always with her. Her behavior is all the more remarkable when we consider that orcas don't have limbs, that a newborn calf weighs hundreds of pounds, and that these animals need to surface regularly to breathe. Only sheer obstinacy could underlie her refusal to let go. The scientists who were monitoring Tahlequah even feared for her health, because due to her obsession with not leaving her baby's corpse behind, she was barely eating enough to recover from giving birth.

Tahlequah's story moved us profoundly. She inspired poems, essays, and hundreds of debates on social media. Outlets like *The Washington Post*, *CNN*, or *The Guardian* informed us of the progress of her sad odyssey.[1] They went even further than this, and when a couple of years later Tahlequah gave birth to a healthy calf, the good news was reported as though a celebrity had just become a mother.[2]

Why did this case become such an obsession for us? Leaving aside the fascination that cetaceans tend to generate in us, Tahlequah's story kept us hooked not just because it was surprising, but because it appeared understandable. We felt we could perfectly comprehend what the whale was going through; we saw in the behavior of this mother a reflection of our own suffering upon the loss of a loved one. This occasion for empathy allowed us to feel momentarily more connected to the animal world.

Nevertheless, some voices called for caution, worried by the idea that we might have been anthropomorphizing Tahlequah's behavior.[3] How could we be so sure that the whale understood that her calf had died? Perhaps she was carrying the body because she believed it was still alive and did not want to leave it behind. And how could we know that the emotion she was experiencing was similar in any way to what we call "grief"? It was more than possible that we were simply projecting onto her our own experiences, our *sapiens* way of seeing the world. When describing Tahlequah in human terms, perhaps we weren't respecting her in her *orca-ness*.

The possibility of anthropomorphizing animals' behavior is a great source of concern for all comparative psychology and ethology, whose experts generally take great care not to attribute to animals more than is strictly necessary to explain their behavior—and especially nothing that is not supported by empirical evidence. Thus, rather than speaking of animal morality, they refer to their "prosociality"; instead of attributing language to them, they attribute "communication"; rather than postulate friendship, scientists talk of "affiliative relationships." This way, scientists protect their disciplines from our almost inevitable tendency to interpret all animal behavior in human terms.

Comparative thanatology, as a subfield of ethology and comparative psychology, was also born with this fear of anthropomorphism. In this case, the danger is all the more evident due to reasons that have to do with its object of study. While other disciplines can set methodological safeguards against anthropomorphism, comparative thanatology, due to its focus on the study of animals' reactions to death, faces ethical limitations that make more difficult the employment of these very safeguards.

In order to understand this better, it's necessary for us to take another little detour to become familiar with the methods of study of animal minds and behavior. Broadly speaking, there are three main ways of approaching the study of how a particular species behaves, and of identifying the cognitive and emotional mechanisms that underlie these behaviors. These are the experimental method, the observational method, and the anecdotal method.

Ways of Studying Animal Minds

The *experimental* method in the study of animal minds consists of subjecting an animal to certain concrete conditions, usually in a laboratory, under which her response to different situations can be put to the test. This method has the advantage of giving us a fair degree of control over the various factors that might be influencing the animal's behavior and, thus, gives us the capacity to test different hypotheses regarding how her mind might work.

For instance, in a series of experiments carried out in the 1990s, Daniel Povinelli and Timothy Eddy wanted to determine whether chimpanzees possess a capacity known as *theory of mind*, which is the ability to understand that other individuals have minds that contain beliefs, desires, perceptions, sensations, etc. More specifically, they examined whether these primates are capable of understanding when someone else can and cannot see them, which would be an example of the capacity to attribute *perception* to others.[4] To do this, they placed chimpanzees in front of two experimenters, to see from whom the apes would beg for food, something they do with a hand gesture—very similar to our own begging motion—with their palm facing up and their fingers extended. What was crucial about the experiment was the fact that one of the experimenters was facing backwards, so that only the other one could see the chimpanzee and, consequently, distinguish her gesture. Under these conditions, the experimenters found that chimpanzees would only beg for food from the experimenter who could actually see them.

However, this result could have been due, not to the fact that the chimpanzee understood that only the experimenter facing forward could *see* her, but to her previous learning of the rule "my gesture works best if others are facing me." In order to distinguish the first hypothesis—the "mentalistic" one—from the second hypothesis—the "behaviorist" one—chimpanzees were subjected to various conditions in which both experimenters were facing them but only one could see them, for the other one

FIGURE 3. Some of the conditions in Povinelli and Eddy's experiment.

was under some circumstance that prevented her from doing so, such as looking upward, wearing an eye mask, or having a bucket over her head (see figure 3). In this case, Povinelli and Eddy saw that chimpanzees would beg for food from both experimenters indiscriminately. Thus, modifying the conditions of the experiment, they were able to distinguish the "mentalistic" from the "behaviorist" hypothesis, and concluded that the chimpanzees' behavior was governed by a behavioral rule and not by an understanding of when others could see them.

Although it gives us the capacity to control variables and test hypotheses, the experimental method has some disadvantages associated with its *ecological validity*. This refers to the fact that the conditions in these experiments are often very artificial and have little in common with the animal's natural environment. In this case, the chimpanzees did not evolve nor were they raised in an environment that included eye masks and plastic buckets, so it's to be expected that this might influence their capacity to distinguish the effect that these objects have on vision. Povinelli and Eddy's experiment was also heavily criticized because in their natural interactions chimpanzees rarely share food with one another, so these experimental conditions might not be ideal for recruiting the social intelligence that is characteristic of these primates.[5]

In later experiments, scientists have opted for the use of competitive rather than cooperative paradigms, as the former are closer to the ecology and social lives of chimpanzees. In one of the most famous studies,[6] two chimpanzees of different status within their hierarchy were given access to an enclosure from opposite sides, and somewhere within that enclosure there was a banana. In some conditions, both primates had visual access to the banana's location, while in others the subordinate chimpanzee was the only one who could see where it was. Experimenters found that the lower-ranking ape was perfectly capable of using information regarding what the other could and could not see in her favor, going for the banana only in those cases in which the dominant chimpanzee had not seen it—and thus sparing herself a beating. Therefore, by bringing the experimental conditions closer to the species' ecology, experimenters obtained results that suggested these animals do in fact have a theory of mind.

(Some people, Povinelli included, weren't convinced by these results, and thought that they could still be explained without appealing to mentalistic reasoning on the part of the apes, but that is another story.)

My point here is that, even though psychologists always attempt to adjust experimental conditions to the ecology and capacities of a species, an animal can fail a test for reasons that have nothing to do with how smart she is (or isn't). It is possible, as in Povinelli and Eddy's experiment, that conditions are just not optimal for the animal to exercise the capacity that scientists intend to test for. Alternatively, perhaps the subject has not understood what is required of her in the experiment; after all, tests don't come with an instruction manual for the animal, and we have no way of communicating with precision what we expect of them. Likewise, perhaps the situation just doesn't generate enough interest in the animal for her to collaborate. After all, the creature doesn't know that she's contributing to the advancement of science (and even if she did, she might not give a damn about it).

An alternative method to this one is the *observational* method, which is characterized by the systematic, long-term monitoring and study of a population of animals that is usually free-ranging. By means of these observations, one can accumulate data about the presence of certain forms of behavior in that population, as well as about the frequency and context in which they occur. Observational studies are closer to descriptions of the natural behaviors of animals and often generate unexpected discoveries. For instance, scientists used to think that the natural use of tools was unique to human beings, until the 1960s, when Jane Goodall began her study of the chimpanzees of Gombe and discovered that their societies exhibit this capacity.[7] Since then, observational studies have documented the use of tools among many different members of the animal kingdom, ranging from primates to insects, and including dolphins, fishes, and octopuses.[8]

Observational studies tend to have high ecological validity, since the animals are normally studied in their natural environment or while they spontaneously interact with conspecifics. However, even though they can tell us a lot about the natural behaviors of animals, these studies have severe limitations when it comes to specifying the capacities that underlie these behaviors, since the fact that there is little control over the environment makes it more likely that unknown factors are influencing what the animal is doing.

Consequently, the observational method is often complemented with experiments performed in the animals' own habitat. In one of the most famous of these, Robert Seyfarth and Dorothy Cheney wanted to test the alarm calls that they had observed in vervet monkeys, for they suspected that these varied depending on the predator that they referred to: leopards, eagles, or snakes. Together with Peter Marler, they carried out an experiment that consisted of hiding speakers in the monkeys' environment in order to systematically study their responses to recordings of different calls in the absence of any predators.[9] This allowed them to confirm their hypothesis, for each call generated a different behavior that corresponded to a defense strategy that the monkeys would adopt when faced with predators of each type: when they played back calls that corresponded to leopards, the monkeys climbed up the nearest trees; when they played back the eagle one, they glanced upward and looked for a bush under which to hide; when the call signaled snakes, they inspected the ground in search of the danger. This experiment confirmed that these monkeys' calls have a semantic content: they refer to something specific in the world and are not a mere expression of their emotional state in the moment.

Observational studies are more useful and reliable the bigger the sample size, that is, the higher the number of individuals being investigated, the number of behaviors documented, and the total time of study of the population. In some cases, however, ethologists can witness an isolated case of a behavior that is particularly surprising, rare, or uncommon, and decide to document it and publish it in a scientific journal. In such a case, we would say that the ethologist is engaging in the third means of studying the animal mind: the *anecdotal* method.

The anecdotal method consists of the opportunistic description of behaviors that have been spontaneously observed in an animal or a group of animals. It's similar to the observational method in that it does not comprise a manipulation of the environment, and instead involves mere observation. In contrast to the observational method, however, it is not born out of an intention to study a concrete phenomenon in a population, but rather consists of reporting a set of behaviors that were observed by chance.

As you can imagine, this is the least reliable method for studying the animal mind, as it offers zero control over the factors that may be influencing the observed behavior, and it favors the radical misinterpretation of it, especially if the behavior has never been seen before. Nevertheless, it's also an important method, for it can open new research avenues or serve to question what we knew until that moment about a particular species.

Psychologists Lucy Bates and Richard Byrne defended a series of criteria that would help to give more credibility to this type of studies. First, they tell us, it's crucial for the described behavior to have been observed by an expert. This is all the more evident now, thanks to the rise of the internet and the subsequent proliferation of viral videos of animals doing bizarre things—things that are time and time again misinterpreted by laypeople.

A particularly good example of this is a video that was recorded in an aquarium, in which we can see a beluga whale interacting with some children who observe her from behind the glass. The whale lunges toward them again and again with an open mouth, as though teasing them, while the kids react with roaring laughter upon each new charge.[10] The person who uploaded the video and the majority of viewers interpreted the whale's behavior like the kids themselves, as though the beluga had been playing peek-a-boo with them. However, cetacean expert Lori Marino argued that the whale was behaving aggressively, and what looked like play was actually threat behavior directed at the children, and probably motivated by a desire to stop their relentless screaming and banging on the glass.[11] It's therefore clear that anecdotes are much more reliable if they come from experts on the animal in question rather than from a WhatsApp group of your old schoolmates.

Second, Bates and Byrne continue, it's crucial for the animal's behavior to be filmed or photographed. If this is not possible, the scientist must note the description of the behavior immediately upon observing it, trying to adopt the most objective terms possible and without adding anything that hasn't actually been witnessed. Similarly, when writing up the corresponding paper, one must not include anything that isn't mentioned in the original notes. This is because human memory is fallible and prone to making stuff up (believe it or not, much of what you remember didn't actually happen that way). This requirement is also important because anecdotes often concern completely unexpected behaviors, so the researcher needs to be alert in order not to have to depend on her memory.

Last, Bates and Byrne argue that, in order to validate the observed behavior, it's necessary that it be confirmed by different researchers on different occasions. An observation carried out by a single ethologist is unreliable and also difficult to interpret, while the same behavior observed by many individuals at different points in time increases the discovery's validity and allows us to obtain more information that can help us to make sense of it. In fact, anecdotal studies are often published as a compilation of observations gathered by different people.

In one of the most famous ones, Andrew Whiten and Richard Byrne presented a compilation of anecdotes that had been collected by different researchers and suggested that primates are capable of intentionally fooling others.[12] These anecdotes included one in which a chimpanzee who was courting a female suddenly hid his erection with his hands upon the appearance of a dominant male; another in which a female gorilla on the march with her family spotted some delicious vine in a tree and stood still, pretending to groom herself, until the rest had passed by and she could have the vine to herself; and one in which a chacma baboon who was being chased by a group of conspecifics who wanted to give him a beating suddenly stood up tall and pretended to look with interest into the horizon, as though there were a potential danger in the area—even though there wasn't—thus managing to distract his aggressors, who forgot what they were doing.

A round of applause, while we're at it, for this baboon.

Lucy Bates and colleagues also carried out an analysis and categorization of all the available anecdotes that suggested that African elephants are capable of empathizing with others and behaving altruistically.[13] Out of all the anecdotes that had been compiled throughout the years, they found seventeen coalition cases, where two or more individuals allied themselves against an unrelated other; twenty-nine cases of protection behavior toward young or injured individuals, including some in which the elephants scared a predator away, intervened to stop a fight among youngsters, or kept them from entering dangerous areas; a hundred and twenty-nine cases of consolation behavior toward a distressed individual; twenty-one cases of caring behavior toward a calf that had been separated from her mother;

twenty-two cases of mothers retrieving a calf that had gotten lost; twenty-eight cases of helping behavior directed at a calf that had difficulties moving; and three cases that involved helping another elephant to deal with a foreign object: a first case in which a male removed a tranquilizing dart from another, a second one in which an adolescent investigated a spear that another elephant had stuck on her back, and a third one in which a mother removed trash from her calf's mouth. These last observations are especially surprising because elephants often carry vegetation on their bodies and have never been seen grooming each other to remove mere weeds.

With anecdote compilations like these, the idea that neither deception nor altruism are exclusive to human beings gained more credibility. They also helped scientists to take more seriously the possibility that animals could engage in these types of behaviors, which led to them being more systematically studied in the lab.

With this brief summary of the methods by which we can study animal minds and behavior, we can begin to understand why comparative thanatology is especially threatened by anthropomorphism.

Comparative Thanatology and the Threat of Anthropomorphism

Because it focuses on the reactions of animals toward death, comparative thanatology is necessarily opportunistic. Animals' deaths in nature for the most part can't be predicted, and there are obvious ethical and environmental considerations that prevent researchers from causing them on purpose to see how the remaining group members react. For this reason, the majority of articles published in this field are reports of anecdotes observed by chance: an individual who falls from a tree and dies, a mother seen carrying her dead infant, an infanticide, etc.

A handful of experimental studies have been performed in this field, and some other designs have been proposed, but in general these pose deep ethical concerns. For instance, Colin Allen and Marc Hauser, inspired by female vervet monkeys' strong emotional bonds toward their babies and their capacity to recognize different individuals through their calls, suggested playing back calls from dead infants through hidden speakers in order to study their mothers' reactions to them.[14] One need only imagine an analogous experiment performed on humans to understand the cruelty of this proposal. André Gonçalves and Dora Biro also suggested a series of experiments based on presenting animals with stuffed subjects that would display ambiguous signs of life and death. For instance, individuals that moved despite smelling of putrescine or having been decapitated.[15] Apart from being quite a sinister idea, it's difficult to predict the psychological impact that these experiments could have on the tested animals.

Observational studies of animals' reactions to death are also difficult, given that the appearance of this phenomenon in nature is too random, which becomes an obstacle for the systematic monitoring of reactions to death in a specific population. Still, some studies of this kind have been carried out in populations with high mortality rates. For instance, Yukimaru Sugiyama and coauthors spent nine years systematically studying the responses of Japanese macaque mothers to the deaths of their offspring, in a population with an infant mortality rate of 21.6 percent.[16] However, this type of research project is rare, for observational studies demand a huge investment of time and resources and are thus commonly devoted to the study of less arbitrary phenomena.

It is therefore safe to say that the vast majority of papers published in the field of comparative thanatology are anecdotal in nature. At the same time, the anecdotal method is the one that most favors anthropomorphism. Since it tends to be based on a sample size of one, and since there are no experimental controls that can help interpret the results, scientists may feel more inclined to interpret the observed behavior through the lens of their human way of understanding the world.

In addition, most studies in this field are not published by people who work exclusively on thanatological issues: they tend to be biologists who are studying a specific population in relation to some topic that has nothing to do with death and who decide to report on a case that they have been lucky enough to witness, like Tahlequah's. There are very few experts in the field of comparative thanatology, based partly on the fact that it is a very young discipline, but also on the clear methodological difficulties to studying animals' behaviors surrounding death. This also increases the chances that the researchers will engage in anthropomorphism, for the less expert the author of a thanatological study is, the greater her tendency to interpret the observed behaviors erroneously.

As though this weren't enough, the risk of anthropomorphism is exacerbated by the fact that death has a great deal of emotional significance for us. It is a phenomenon that we have a hard time dealing with objectively, because we associate it with very traumatic personal experiences or because it connects with our most existential fears. (I know this well, for I have killed the mood of more than one dinner party by discussing my research on these topics.) Consequently, there exists the threat of us projecting onto animals our own feelings and not being able to avoid seeing human emotions in their behavior, as in the case of Tahlequah.

Given all these risks, the thanatological behaviors of animals are often reported in the most restrained way possible and avoiding the use of terms, such as *grief* or *concept of death*, that hold strong human connotations. This way, researchers aim to compensate any possible anthropomorphic bias that may interfere with the correct interpretation of the animals' behavior. Even though taking care not to fall prey to anthropomorphism is healthy and important when it comes to studying animal minds, there are two biases that are also dangerous but that elicit much less attention. These are *anthropectomy* and *anthropocentrism*. Let's have a closer look at what anthropomorphism consists of and what distinguishes it from these two other phenomena.

Anthropomorphism, Anthropectomy, and Anthropocentrism

In strict terms, anthropomorphism is just the attribution of human qualities to something or someone. Thus, we can describe the Christian god as an anthropomorphic god, for he has a human form (even though the religious doctrine postulates this the other way around, establishing that we are the ones shaped like gods). The famous faces of Bélmez* are also anthropomorphic figures. In its most literal etymological sense, anthropomorphism is neither good nor bad; it simply consists of assigning a human quality to a specific entity.

However, in comparative psychology and in ethology, that is not how this term is typically used. When a scientist criticizes us for anthropomorphizing an animal, she's not just implying

---

* A supposedly paranormal phenomenon in which face-shaped stains mysteriously appeared on the floors of a house in rural Spain.

that we're interpreting her behavior in human terms, but also that we're making a mistake. Anthropomorphizing Tahlequah is not just attributing human qualities to her, but attributing human qualities *that she lacks*.

Even though this is the common use of the term in these disciplines, it's a somewhat problematic notion, for it suggests that all human qualities are exclusively human. But, of course, we know that this is not the case. Not just because we obviously share many morphological and physiological characteristics with other animals (like our spine, organs like the heart or lungs, hormones, etc.), but also because many psychological capacities that used to be considered uniquely human, like tool use, deception, or altruism, have been uncovered in many other species.

Given that in the past we have often erred when assuming that certain capacities were exclusively human, philosophers Kristin Andrews and Brian Huss believed it was necessary to coin a term that captured that error and that would thus serve to complement the notion of anthropomorphism. They suggested the term *anthropectomy* (from the Greek words *ánthropos*, meaning "human," and *ektomía*, "to cut out").[17] Anthropectomy would be the other side of the coin with respect to anthropomorphism. If anthropomorphism is the mistaken attribution of a human-typical characteristic to an animal, anthropectomy would be the mistaken denial of a human-typical characteristic to an animal.

What is crucial about the term that Andrews and Huss coined is that anthropectomy would be an *equally grave* error as anthropomorphism. Thus, if we attribute a concept of death to Tahlequah or we say that she is grieving, and it turns out that we are wrong, then we are anthropomorphizing Tahlequah. But if we say that Tahlequah lacks a concept of death or is not grieving because these are uniquely human capacities, and it turns out that we are wrong, then we are anthropectomizing Tahlequah. Both mistakes are equally serious, and there is no reason to fear one over the other. They are both false descriptions of reality. Therefore, even though comparative thanatologists do well to fear anthropomorphism, they should also beware of anthropectomy.

Anthropomorphism also needs to be distinguished from *anthropocentrism*. In its most general sense, anthropocentrism refers to a bias that leads us to consider our own species as the center of the universe. It's the tendency to think that, given that we are the most important thing from our perspective, we must be the most important thing in general.

Anthropocentrism can manifest in different ways depending on the context. Most of our cities, for instance, follow an anthropocentric urban planning, for they systematically ignore the many nonhuman species with whom we share a space; not just our pets (which are for the most part not massively taken into account either), but the thousands of wild species that inhabit and are adapted to the urban environment, as is the case of rats, cockroaches, pigeons, sparrows, squirrels, racoons, etc.

In science, anthropocentrism also manifests in various ways. Although according to its popular image science offers objective facts about the world, the truth is that this is an activity that is carried out by humans, and as such it is also a target for our biases and is heavily influenced by our values. The topics that we decide to study, the ones that have the greater chances of getting published or receiving funding, are not random, but rather obey our human interests. For this reason, the relevance of studying the mind and behavior of animals is often justified with reference to some human value, like our interest in uncovering the evolutionary origin of our psychological capacities or what distinguishes our mind from that of the remaining species.

Comparative thanatology cannot escape the influence of our human values either. In this discipline, the study of animals' behavior and reactions surrounding death is often justified by an appeal to the importance of discovering the evolutionary origin of our own funerary practices and attitudes toward death. Even though this is of legitimate interest, it's important to point out that a question doesn't necessarily become more interesting when it can be explicitly related to us and our values. Perhaps you, the person reading this, find the question of how animals experience and understand death fascinating in and of itself, independently of what it can tell us about ourselves. If that is the case, welcome to the club.

In addition, we must remember that the study of other animals' behavior doesn't have to give us clues about the evolution of our own capacities, even in the case of closely related species, such as the other great apes. Chimpanzees, bonobos, orangutans, and gorillas are not "less evolved" humans. They are species with whom we share common ancestors, but at no point in our evolutionary past were we chimpanzees, bonobos, orangutans, or gorillas. Consequently, even though studying them can give us some answers about our own evolution, they are not a direct window into our evolutionary history.

Anthropocentrism is a bias that leads us to believe that we are the center of the universe, but it also works like a pair of innate blinkers that prevent us from seeing beyond our own perspective. This anthropocentrism is implicit in a great deal of studies in comparative psychology, which don't just aim to answer questions with our interests in mind, but also often follow methodologies that highlight the difficulties we have in abandoning our *sapiens* way of seeing the world.

Let's take as an example the famous mirror self-recognition test. This experiment was designed by Gordon Gallup and is used to determine whether animals have the capacity to recognize themselves, which would be evidence of a form of self-awareness.[18] In the original experiment, captive chimpanzees were isolated from other group members and placed in a cage inside an enclosure on their own for two days. After this, a mirror was placed in the enclosure and the chimpanzees' reactions observed. The apes first reacted to their reflections as though they were other chimpanzees and carried out social behaviors like vocalizing or displaying. However, as the days went by, these social behaviors disappeared and the chimpanzees began to behave as though they were recognizing themselves in the mirror, using it to inspect or groom areas of the body that they usually don't have visual access to, like their nose, the inside of their mouth, or their anus.

In order to test the hypothesis that they were recognizing themselves, Gallup decided to go one step further. The chimpanzees were anaesthetized and a red dye was used to place an odorless mark on their forehead and behind their ear. Before they woke up from anaesthesia, the mirror was removed from the enclosure. When the chimpanzees awoke, they didn't notice anything strange and went about their day as usual—something that radically changed when the mirror was reintroduced into the room. In that moment, they began to insistently touch their marks, as though they were asking themselves what in the world that thing on their face was, which would prove that they were indeed recognizing themselves in the mirror. Gallup interpreted these results as evidence of a form of self-awareness in this species.

Although passing the mirror test shows us that the subject has learned how to use a mirror—which is, we might add, not a trivial task—and that she possesses at least the capacity to recognize her body, an animal can fail this test for reasons that have nothing to do with a lack of self-awareness, but rather with the fact that they are not humans. For instance, gorillas tend to fail this test, which has been linked to the fact that, unlike us, members of this species tend to avoid eye contact with conspecifics, as it is considered a threat behavior.[19] For this reason, they might not pay sufficient attention to their mirror image to realize that it corresponds to their own movements.

Other species may fail this test because they are not predominantly visual animals like we are. For instance, dogs fail the mirror test but pass an analogous test based on olfaction.[20] In this experiment, dogs were presented with urine samples corresponding to themselves and to other dogs, and they were found to spend less time smelling their own samples. However, if a weird olfactory "mark" was added to their own urine, their interest in it skyrocketed. This suggests that they can discriminate their own smell, which would also be a form of self-recognition.

Despite these problems, the mirror test is usually considered the gold standard in the study of animal self-awareness, which evidences an anthropocentric perspective that makes it difficult for us to appreciate that other animals can recognize themselves primarily through other senses. Not only this—the mirror test also presupposes that the tested animals *care* about their appearance in one way or another, for something like this is necessary for them to have the motivation to interact with the mark that has been placed on them. However, it's not immediately clear that other animals are vain enough to care about their own looks. Whether or not they do care will be related to their evolutionary history and their particular ecology.

For example, in a recent study, a species of fish, the cleaner wrasse, was found to be capable of passing the mirror test.[21]

Although one in principle would not expect something like this from a tiny little fish, when one learns how this species makes a living, these results appear much less surprising. These fishes feed off the parasites that they visually detect on the skin of other fishes, and the mark that was put on their throat mimicked the size, color, and shape of one of these parasites. When there was no mirror, the little fishes would ignore the mark, but when the mirror was brought back in they immediately started to try to remove the mark by scraping their bodies against the substrate. In the world of these fishes, this mark is something worthy of attention, but this does not have to be the case for all animals. For others, perhaps an unexpected smell in their pee or their fur is much more interesting and attention-grabbing than having a little grime on their forehead. The fact that we struggle to understand this is only proof of the anthropocentric glasses with which we look at the world.

In comparative thanatology, anthropocentrism manifests in a similar way, through the adoption of human experience as the gold standard against which we compare all animal behavior around death. Thus, we can distinguish on the one hand an *intellectual anthropocentrism*, which is the assumption that the only way of understanding death is the human way, and on the other hand an *emotional anthropocentrism*, which consists of presupposing that the only way to emotionally react to death is the human way.[22] In the next two chapters, I will discuss in detail these two forms of anthropocentrism and defend a view of what we can do to ensure that we leave them aside.

We have now seen the potential biases that threaten comparative thanatology. Given its methodology and its object of study, this discipline is especially prone to anthropomorphism. However, the obsession with avoiding this bias can increase the chances of falling prey to anthropectomy, and there is no reason to fear this bias less than anthropomorphism. Similarly, anthropocentrism also constitutes a grave risk, for it is a subtle bias that may however greatly influence the methods of study and the phenomena that are considered to be of scientific interest. Therefore, the same care that scientists usually take to avoid anthropomorphism must also be put into avoiding anthropectomy and anthropocentrism. When it comes to interpreting cases like Tahlequah's, we must be sure not to attribute more than is warranted, but also not less than we should.

4

The Ape Who Played House with Corpses

In the tropical jungles of Kibale National Park, in Uganda, lives a population of chimpanzees that has been systematically and continuously studied since 1995. Despite all these years of research, in 2016, Jacob Negrey and Kevin Langergraber came across a behavior that had never been seen before.[1] An adult female chimpanzee called Lucy was going about her day-to-day activities while carrying the corpse of a bush baby (a nocturnal primate the size of a squirrel that has gigantic eyes and ears; see figure 4). When she ate, she used both hands to manipulate the fruit and held the little body between her abdomen and her thigh. When taking a nap, they saw how she made a nest and proceeded to rest next to the carcass. When she moved about, she carried the body on her back. They also observed her carefully grooming it. When one of the youngsters of the group approached with curiosity, Lucy prevented them, calmly but firmly, from getting close to her "treasure." Lucy's interaction with the corpse lasted for five and a half hours, during which time she traveled with her group for more than one kilometer.

FIGURE 4. Lucy grooming and carrying the bush baby's corpse.

Various things are noteworthy about this case. The first of these is the fact that chimpanzees tend to consider bush babies as a precious delicatessen item, and yet Lucy displayed no interest in eating this particular specimen. The second point is that the behaviors that Lucy was directing at the carcass correspond to maternal care in this species. When their babies are very young, chimpanzee mothers keep them close at all times, even when they eat or sleep. They also groom them, carry them on their backs, and make sure that the other group members' curiosity doesn't pose a danger to them. Lucy appeared to be playing house.

Lucy's is not the first case of play-parenting observed in chimpanzees. As primatologists Sonya Kahlenberg and Richard Wrangham point out, it's fairly common to see young female chimpanzees carrying sticks or logs in baby-like ways, holding on to them for minutes or hours at a time while they travel, eat, rest, or sleep.[2] The fact that this behavior is mostly seen in females, that it disappears after their first pregnancy, and that it shares characteristics with maternal care suggests that these chimpanzees are role-playing motherhood; perhaps emulating the behavior that they see in the adults and preparing for when their own time comes.

What is special about Lucy's case is that her behavior was not directed at a stick, as is usually the case when chimpanzees play-parent, but rather at the corpse of an individual of another species. Moreover, it's also worth highlighting that Lucy was far from young: at twenty years of age, she was already a fully grown adult. This game is for youngsters, not for grown-up chimpanzees. However, Lucy is different from other females in her group in that she appears to be sterile. She is not known to have given birth, despite the fact that these primates tend to have had their first baby by the time they reach thirteen. It's more than likely that this lack of offspring influenced her behavior and perhaps also led her to direct it, not at a piece of wood, but at an entity that more closely resembled a baby of her own species.

In fact, this was not the only time that Lucy was seen carrying out this behavior. One year later, she was observed playing house again, this time with the carcass of an infant chimpanzee. It was the son of Bronte, another female from the same group.

The little one had died when he was five months old, and the mother had been carrying the corpse for a couple of days (a behavior that reminds us of Tahlequah's and that, as we shall see in the following chapter, is common among primate mothers). Bronte was sitting in a tree having a snack when Lucy walked up to her and snatched the carcass. Bronte shrieked and started chasing Lucy but gave up after a few meters. During the next two hours, Lucy carried the corpse as though it were a living baby and spent several minutes grooming it. She was also seen slapping and shaking it. When she at last grew tired, she dropped the body and continued marching with the group. In that moment, Bronte walked up and sat on the ground next to the carcass. She observed it for a minute while swatting away the flies that were starting to circle it. When she noticed that the group was leaving her behind, she grabbed the body and started following the rest. However, fifteen minutes later she dropped it and never again showed any interest it.

Bracketing the fascinating issue of the reason behind Lucy's interest in pretending to be a mother, the inevitable question emerges: did Lucy know that her "babies" were dead? And at the same time, had Bronte understood that her baby was no longer alive, and was that why she showed so little motivation in getting him back? Did Lucy and Bronte have a concept of death?

Philosophers have a tendency, to the annoyance of their friends, families, and students, to answer all questions that are posed to us with an "it depends." This stems from the importance that philosophy gives to conceptual characterizations. If you've spent enough time around philosophers, you will already have anticipated that the answer to the question of whether Lucy and Bronte had a concept of death *depends* on what we understand by that notion. Indeed, for us to answer this question we need to begin by giving a clear account of the meaning of this term. Now, not all characterizations are the same. Some carry with them undesirable biases. And in fact, this has been the case for the way the concept of death has usually been understood in debates on comparative thanatology. This term has, more often than not, been characterized in a way that evidences intellectual anthropocentrism, a common and very problematic bias that I already introduced in the previous chapter and that we can now examine more closely.

Intellectual Anthropocentrism

If we were to take the *sapiens* way of understanding death as the standard by which to determine whether animals possess a concept of death, we would be engaging in intellectual anthropocentrism. That is, this is a bias that consists of defining the concept of death from a purely human perspective. It generates the implicit assumption that we will only be able to say that animals have a concept of death if they have *our* concept of death. But of course, our concept of death is extremely complex. With that as their reference point, scholars have postulated demanding cognitive requisites for understanding death, such as the capacities for analogical reasoning, abstract thought, and a theory of mind, in addition to concepts such as those of life, time, or absence.[3]

As we shall see, this amounts to an overintellectualization of the concept of death. At the same time, it almost inevitably generates the conclusion that the concept of death is very difficult to acquire and that it's going to be a very rare feat once we move beyond our own species. For instance, despite the fact that there are many thanatological studies on monkeys, and that we know that these animals show a wide range of reactions to death, Arianna de Marco and collaborators recently stated that "there is no reason to believe that monkeys have a concept of death."[4] Perhaps it's true that monkeys don't have a concept of death, but, given the available evidence, stating that there is *no reason* to believe the contrary can only be sustained if we are presupposing a highly intellectualistic notion of the concept of death; a notion that makes it practically impossible for nonhuman animals to acquire it.

Similarly, André Gonçalves and Susana Carvalho, after a detailed review of all the thanatological studies performed on primates, conclude that only great apes appear to be good candidates to acquire a concept of death similar to humans'. This is once again evidence that the human concept is taken as the reference point and that it is presupposed to require a lot of cognitive complexity.[5]

Scientists are not the only ones who are guilty of this intellectual anthropocentrism about the concept of death. In fact, the majority of philosophers who have discussed the topic have also fallen prey to this bias.

In animal ethics, there is a debate as to whether death constitutes a harm to the animal who undergoes it. Even though practically all philosophers agree that pain harms animals, in the case of death—and assuming that it's quick and painless—opinions are much more divided. In particular, some defend that, for death itself to be a harm to someone, one needs to understand what death is. Therefore, the fact that animals supposedly lack a concept of death would imply that death in itself doesn't harm them.

The philosopher Ruth Cigman, for instance, argued that having a concept of death is necessary for the possession of "categorical desires."[6] This pompous-sounding term refers to those desires that serve to answer the question "why do you want to stay alive?" They are the reasons why we want our life to continue; things such as the desire to see our children grow or to finish our masterpiece. Having this kind of desire is necessary, according to Cigman, for death to harm us. If we lack a concept of death, we can't have these desires, and therefore death would not do us any wrong.

In a similar vein, another philosopher, Bernard Rollin, argues that only those things that explicitly matter to us can harm us. At the same time, for death to matter to us we need to have a concept of death.[7] If you're incapable of understanding that you can die, death can't be something that matters to you. Similarly, Christopher Belshaw has argued that, for death to be harmful, we need to possess the desire to go on living.[8] Death would harm us because it would frustrate this desire. In turn, possessing a desire like this requires understanding the difference between life and death, and thus, requires a concept of death.

Naturally, not all philosophers agree with these arguments. Ben Bradley, for instance, has pointed out that one does not need to understand that something is a harm for it to be so.[9] An individual may lack a concept of cancer and not understand that cancer can harm her, but this lack of understanding clearly won't prevent her from being harmed by it. Elizabeth Harman, in line with many other philosophers, has argued that death harms us because it deprives us of those goods that we could have enjoyed in the future.[10] Therefore, for death to harm us, a concept of death is not needed, just the capacity to enjoy different goods, such as love or friendship. And for Tom Regan, death would be a harm whenever an entity with a psychological life and inherent value ceases to exist with it.[11] For this, once again, a concept of death is not needed.

Despite such disagreements, the vast majority of philosophers who take part in this debate accept the other premise that is implicit in the original argument: the idea that animals lack a concept of death. This is quite surprising, given that philosophers very rarely agree with each other. Whenever one finds an issue on which all philosophers agree, there's usually something fishy going on, for if there's one thing that characterizes philosophers, it's our passion for arguing with one another. And indeed, if one submerges oneself in the literature in search of an argument to support this premise, what one will generally find is that it's offered with a total absence of empirical evidence to back it up. Yet it's accepted without complaints by all participants in the debate. For instance, Cigman simply asserts: "It is only by an imaginative leap that possession of [the concept of death] seems attributable to animals,"[12] and Regan, who extensively criticizes Cigman, nevertheless agrees with her that "[i]t is doubtful that any animals can [understand their own mortality]."[13]

In the few cases in which this premise is defended with arguments, what once again emerges is an intellectual anthropocentrism at the root of philosophers' confidence in the idea that animals lack a concept of death. Rollin, for instance, invoking the German philosopher Martin Heidegger, asserts that:

> to understand death requires that one grasp the "possibility of the impossibility of one's being," a notion involving a possible state of affairs, which, as a counterfactual, requires sophisticated syntax that there is no reason to believe animals possess.[14]

Rollin is not just linking the possession of a concept of death to the capacity for counterfactual reasoning—something already very complex—but is also taking as his point of departure an idea of Heidegger's, a thinker who is famous in the philosophy classrooms of any university for his extremely cryptic theories. Rollin is defining the concept of death in a way that is *necessarily* going to leave out all nonlinguistic animals, and perhaps also many who do possess a language. In fact, I'm sure that many humans to whom we would attribute a concept of death without giving it a second thought would be incapable of explaining to us what on earth is meant by the expression "the possibility of the impossibility of one's being."

This intellectual anthropocentrism not only leads to an over-intellectualization of the concept of death, but also implies the presupposition that the concept of death is something binary: something that you either have or you don't, but that doesn't admit any kind of gradation. But this doesn't correspond to how we think of the concept of death in the case of humans' psychological development. When we think of how kids learn about death, we're perfectly capable of understanding that there's a scale here. Babies are not born with a concept of death, nor do kids acquire it overnight. In fact, until they are about ten years old, many children lack a fully formed concept of death,[15] but anyone who's played Super Mario Bros. with a six- or seven-year-old will know that at that age they already have some notion of death, as did the little girl with the microscope that I told you about in the second chapter.

We could thus be accused of applying a double standard. In the case of human children, we admit a gradation. In the case of animals, we assert: either they have a concept of death equivalent to the average adult human's or they lack a concept of death. All or nothing, yes or no, black or white, pizza with pineapple or sanity.

Not only is this an unfair way of posing the question, it's also rather arbitrary to think of the average adult human's concept of death as the "all." Can we really assert that adult humans have a *full* concept of death? Not only does death equate to a huge mystery for many people, given that we don't know what comes after it, but even for those of us who are convinced that death is the annihilation of our consciousness, this appears not completely conceivable due to limitations of our own intellect. This was marvelously put by the Spanish philosopher Miguel de Unamuno:

> Try, reader, to imagine to yourself, when you are wide awake, the condition of your soul when you are in a deep sleep; try

to fill your consciousness with the representation of no-consciousness, and you will see the impossibility of it. The effort to comprehend it causes the most tormenting dizziness. We cannot conceive ourselves as not existing.[16]

If we must think of it in strict all-or-nothing terms, why not think of the "all" as including incontrovertible knowledge of what happens after death, or the capacity to "conceive ourselves as not existing"? Under these conditions, all animals on the planet, including humans, would lack a concept of death.

Perhaps some reader is convinced by this binary way of seeing things; still I doubt that it's a practical route to follow. I believe it's better to think of the concept of death as a spectrum, one in which human beings' concept of death would lie to one side of the center, even if not right at the end. The question would then become: where do the remaining animals fit?

This way of understanding the question also makes it, in my opinion, much more exciting than how it has normally been formulated. If the question is, "do animals have a concept of death equivalent to the typical adult human's?," the answer is very obvious: no. This not only would have made for a very short book, but there's also nothing interesting or surprising about this answer: it's self-evident that animals that lack the narratives surrounding death that we have or a cumulative culture of traditions related to mortality aren't going to have a concept of death that is as complex or as sophisticated as ours.

The interesting question, the one that leaves room for discussion, is rather whether animals have *anything that counts* as a concept of death and, if so, how complex this concept is and what can or can't they do with it. This is the question that I will try to answer in the remainder of this book.

In order to answer the question of whether animals have anything that counts as a concept of death, we must first establish the minimal conditions that an animal needs to fulfill for it to be legitimate to attribute to her such a thing: this is what I will refer to as the *minimal concept of death*.[17] This is necessary in order not to put the cart before the horse, as has been done until now in the debate. Instead of beginning from human beings' hypercomplex concept of death and asking if animals have that, we shall begin by asking whether they fulfill these minimal requirements and then inquiring into how complex animals' concept of death can become.

The Minimal Concept of Death

At this point it's crucial for us to recall the conditions we established for general concept possession, given that taking care not to require too little is just as important as preventing an over-intellectualization. If our bar is set too low, we run the risk of turning the minimal concept of death into something trivial that has little to do with what we have in mind when we attribute an understanding of death to an individual. An excessively liberal conception of the minimal concept of death will thus be just as undesirable as an excessively conservative one.

For the minimal concept of death to be a *concept*, it has to fulfill the six conditions I set out in the second chapter: (1) it has to allow the animal who possesses it to distinguish dead entities with some degree of reliability, (2) it has to come with a certain understanding of what philosophers of language call the semantic content of the property of being dead, (3) it has to leave room for a variation across cultures, species, and individuals, (4) it has to allow for inferences to be made with it,

(5) it can't be linked to a concrete sensory stimulus, and (6) it mustn't generate a fixed behavioral response. This way, we will ensure that animals that are only capable of stereotypical reactions to death won't count as possessing a concept of death.

In order to characterize the minimal concept of death, I will begin by focusing on condition (2). We need the animal to have a certain understanding of the semantic content of the property of being dead. That is, it's necessary for the animal to have a minimal comprehension of what it means for someone to die. But just how much knowledge is necessary to say that an animal understands death?

A good way to start delineating this are the seven subcomponents of the concept of death that developmental psychologists use to determine to what extent children of different ages understand death.[18] These seven subcomponents are:

1. *Non-functionality*: death implies the cessation of bodily and mental functions.
2. *Irreversibility*: Dead individuals can't come back to life.
3. *Universality*: All living individuals, and only living individuals, die.
4. *Personal mortality*: Death will also apply to oneself.
5. *Inevitability*: Death can't be indefinitely postponed.
6. *Causality*: Death is caused by a breakdown in vital functions.
7. *Unpredictability*: The exact time of death can't be predicted in advance.

These seven subcomponents correspond to a scientific or biological view of death. In some human cultures, especially those that postulate the existence of supernatural entities, some of these subcomponents may not be present or may just be partially present. For instance, in some religions it's considered that some living beings don't die, which would imply that these cultures don't incorporate the subcomponent *universality*. Others consider that just bodily functions, but not mental functions, cease upon death, so they would only present a partial version of the *non-functionality* component. And still others incorporate subcomponents that are not among these seven, like the idea that after death one goes to heaven or hell, or reincarnates into one or another being depending on how one behaved in life.

Despite these cultural differences, we can speak of a series of ideas common to all concepts of death proper and that must be present for it to make sense to say of an individual—be it human, animal, or alien—that she possesses a concept of death. To determine what these are, let's have a closer look at these seven subcomponents.

We begin with the subcomponent *non-functionality*, the idea that upon death all mental and bodily functions stop. As I've mentioned, according to certain nonscientific conceptions, some of these functions can survive death. However, it would be strange to say of a Christian—someone who believes that her mental functions will continue existing after her death—that she lacks a concept of death or has a *radically* different one from the scientific concept of death. A Christian and an atheist can still perfectly communicate with each other about death, as is common in those societies in which religious and secular people coexist. An anarcho-punk teenager can perfectly understand his Catholic grandmother when she tells him that Torcuato died, even though the grandmother thinks Torcuato has reunited with his deceased parents in heaven and the teenager thinks that he's feeding the maggots and little else.

However, let's imagine an individual called Pelayo who thinks that *all* functions, both mental and bodily, survive death, that is, he thinks that after you die you can still feel sadness and joy, laugh and cry, eat, dance, and everything else you can do with the mind and the body. Both the Catholic grandmother and the anarcho-punk teenager will think that Pelayo is gravely confused about what it means to be dead. If Pelayo told them that he's going to a library to get himself some pants, they would think that Pelayo doesn't understand what a library is. Similarly, when hearing him talk this way about dead people, they would think that Pelayo *doesn't understand what it means to be dead*, that he lacks a concept of death.

Therefore, a certain degree of non-functionality must be present in a minimal concept of death. The functions that one considers to cease with death, in turn, will correspond to those that the individual considers to be *characteristic* of living beings. For the Catholic grandmother, these are just the bodily functions, for under the Christian conception usually only the soul is considered immortal and so mental functions characterize both living and dead individuals. For atheists like the anarcho-punk teenager, in contrast, both bodily and mental functions cease upon death.

For us to be able to assert that Pelayo lacks a concept of death, the minimal concept of death has to incorporate, to a certain degree, the subcomponent *non-functionality*. At the same time, for it to make sense to say that both the anarcho-punk teenager and his Catholic grandmother have a concept of death, the minimal concept of death must allow different conceptions of the functions characteristic of living beings. This is going to be very important in the case of animals. By not fixing ahead of time the group of functions that an individual has to understand as ceasing with death, we will be able to incorporate some variability in the concept of death across species, across individuals, and throughout time.

This variability is crucial, because any notion of nonfunctionality that we can find in an animal will reflect *how that animal understands the functions of the living*. This should be obvious enough, for we can't expect an animal to understand that dead individuals lack functions that the animal is unaware that living individuals have to begin with. Thus, for example, we can't expect a rabbit to understand that death implies the cessation of digestion, because rabbits—we can assume—lack a notion of digestion.

This, which should be as plain as the nose on one's face, hasn't always been taken into account, for a theory of mind has sometimes been postulated as a prerequisite for the concept of death.[19] As I mentioned in the previous chapter, a theory of mind is the capacity to attribute mental states to others. This ability would thus be necessary for an animal to understand that individuals lose their mental functions when they die. Given that the presence of a theory of mind in the animal kingdom is doubtful, even in the case of chimpanzees,[20] this is seen by some thanatologists as a reason against the possibility that animals have a concept of death. However, just as in the case of digestion, it only makes sense to expect a concept of death to incorporate an understanding of the cessation of mental functions in those cases in which the animal *in fact* possesses a theory of mind (not to mention that requiring this would imply that the Catholic grandmother would lack a concept of death).

We can see this more clearly with a thought experiment of the sort that philosophers enjoy so much. Let's imagine that in the future scientists discover a physiological process that characterizes the human body when it's alive, and let's call it *jepperfication*. Let's suppose that, for some arbitrary and mysterious reason, we didn't know anything about *jepperfication* before it was discovered. Clearly, it would make no sense to say that before we knew about *jepperfication* we lacked a concept of death because we didn't know that with death came the cessation of *jepperfication*. Well, for those animals who lack a theory of mind, others' mental states are equivalent to *jepperfication*. Therefore, it's absurd to establish that they can't have a concept of death if they lack a theory of mind.

If an animal has a notion of non-functionality, this will mirror her notion of functionality, with whichever limitations it comes. This also means that it will incorporate her particular way of seeing the world, as it did in the case of the Catholic grandmother and her anarcho-punk grandson. An animal with a minimal concept of death will understand that dead individuals don't do the things that *she understands as characteristic of living beings*, which might include, for example, moving, vocalizing, eating, playing, or mating.

This allows us to incorporate a variability in the concept of death at the *species* level. Thus, for instance, for species with a keen sense of smell, dead individuals may be fundamentally characterized by the fact that they stop smelling a particular way; while for species whose members are very competitive, dead beings may be mainly characterized by not competing for available resources. Similarly, this allows us to incorporate a variability at the *individual* level. For instance, individuals with very playful personalities may consider that what is most noteworthy about the dead is that they don't play. It also allows for a variability across *time*, such that, as the individual matures or acquires more knowledge, her notion of the living's functionality may change.

A minimal concept of death thus needs a certain degree of understanding of the subcomponent *non-functionality*, though we can think of this as a notion that allows for a lot of variability, depending on how subjects understand the functionality of the living.

Let's now consider the next subcomponent: *irreversibility*. This is the idea that the dead can't come back to life. When children start learning about death, they tend to see it as something reversible, such as falling asleep or getting sick.[21] The minimal concept of death has to capture why we consider this to be a pretty flagrant mistake. At the same time, we want the minimal concept of death to encompass death conceptions that incorporate some element of reversibility, such as the notion of reincarnation.

The best way to solve this, in my view, is to incorporate the notion of irreversibility but connect it to the group of functions that one considers to cease with death. This way, an individual who believes in reincarnation will continue having a concept of death, for reincarnation occurs in a different body and with no memories of the past life, which means that certain bodily and mental functions that stopped with death are not recovered.*

This notion allows us to say that someone who believes in reincarnation still has a concept of death, while good old Pelayo,

---

* Some Christians believe that after death the body will be resurrected together with the soul. This could be taken to mean that they consider there to be complete reversibility. However, while I'm not an expert on Christian dogma, I would assume that this view doesn't entail that bodies are resurrected with all their functions. Otherwise, if I were to die now I would have to experience period cramps and regularly trim my toenails in heaven, which hardly sounds like paradise. So presumably at least *some* bodily functions are lost irreversibly under this conception. Even if I'm wrong, though, it's not a big deal. The argument I will build in the chapters to come works just as well if we consider irreversibility and non-functionality to be *sufficient* if not necessary components of a concept of death.

who has finally understood that the dead don't do things, but now thinks of death as completely reversible, still doesn't have a concept of death. At the same time, this notion allows us to assert that an animal who thinks that this dead body is not doing anything right now, but will get up and start doing stuff any time soon, is an animal who has not understood what it means to be dead and is operating with a concept closer to that of being asleep.

The following three subcomponents, *universality*, *personal mortality*, and *inevitability*, can be seen as a whole, for if *universality* is true—that is, if *all* living beings and *only* living beings die—then *personal mortality* and *inevitability* will be entailed. None of these three subcomponents is necessary for a minimal concept of death. This is because in principle it's possible for an individual to understand death as something that occurs only to *some* individuals and still have a concept of death.

We can illustrate this with another thought experiment. Let's imagine that in the future, and due perhaps to the felicitous discovery of *jepperfication*, scientists inform us that not all humans are going to die, or that we ourselves are not going to die, or that there is some way to postpone death forever. This would not fundamentally alter our concept of death, though it would undoubtedly change how we live our lives. In this context, if we found out that poor Torcuato died, this would still mean essentially the same: that Torcuato has lost the group of functions characteristic of living beings and will remain forever in that state. Although in this imaginary world there would be an extra dimension of tragedy to Torcuato's death, we can still make sense of it with our ordinary concept of death. Therefore, *universality*, *personal mortality*, and *inevitability* are not necessary for a minimal concept of death.

An understanding of the next subcomponent, *causality*, would undoubtedly be advantageous for any animal. Understanding the causality behind death would increase the probability of her surviving until reproduction, by allowing her to avoid potential sources of danger. However, a minimal concept of death also doesn't require the subcomponent *causality*, for the simple reason that, in general, having a concept of X doesn't require understanding the causes of X. Thus, for instance, one can have a concept of arthritis without knowing what causes arthritis. Learning about what causes arthritis would make our concept more complex and perhaps more useful, but the concept of arthritis itself doesn't require it. Therefore, we can also relinquish the subcomponent *causality*, despite the evolutionary advantages that may come with it.

With respect to the last subcomponent, *unpredictability*, the first thing we need to say is that it's not strictly speaking true that death is intrinsically unpredictable. In certain contexts, such as the execution of someone with a guillotine, the exact timing of death can be predicted pretty reliably. This doesn't make their death mean something different. In fact, if we learned that hard determinism is true (that is, if we knew for a fact that free will doesn't exist and everything is predetermined) and we were capable of calculating the exact moment of our deaths, this would also not fundamentally change what it means to die. It would for sure imply a radical change in the way we live our lives, but dying would still mean the irreversible cessation of those functions that characterize living beings. Therefore, the subcomponent *unpredictability* is also not necessary for a minimal concept of death.

The crucial subcomponents for a minimal concept of death are, therefore, *non-functionality* and *irreversibility*. In more technical terms, we can define the minimal concept of death as follows:

> An animal possesses a minimal concept of death once she can classify, with a certain degree of reliability, some dead individuals as dead, where *dead* is understood as a property that pertains to beings who:
>
> (a) are expected to have the cluster of functions characteristic of living beings of their kind, but
> (b) lack the cluster of functions characteristic of living beings of their kind, and
> (c) will not recover the cluster of functions characteristic of living beings of their kind.

I have written it in this pedantic manner for two reasons. First, because I want the minimal definition to be well delineated, for in the chapters that follow I will return again and again to this notion of a minimal concept of death. Second, to be able to highlight the reasons why the minimal concept of death is indeed a concept. Ready? Let's go.

Why the Minimal Concept of Death Is a Concept

The minimal concept of death, first, allows the animal who possesses it to distinguish dead entities with a certain degree of reliability. It's not necessary for the animal to be able to discriminate the death of *any* creature, for the subcomponent *universality* has been cast aside. It's also not necessary for the animal to be 100 percent reliable in her discrimination of death. It can sometimes happen that she takes an individual who is asleep or in a coma for dead, no big deal. But if she were to make a mistake in her classification *every single time*, then it wouldn't make sense to say that she possesses a minimal concept of death. So, a certain degree of reliability is necessary, but not total reliability.

Second, the minimal concept of death comes with some understanding of the semantic content of the property of being dead. That is to say, the animal understands to a certain extent what it means to be dead. In particular, the animal understands that dead individuals don't do the things that living individuals usually do (*non-functionality*), and that this is a permanent state (*irreversibility*).

In addition, the minimal concept of death presupposes a minimal concept of life, which is determined by condition (a) and establishes that the animal has to have certain expectations about how living beings of that particular kind usually behave. This is important because, as we saw, every concept must be in a semantic net with other concepts and needs the latter in order to acquire meaning. Without a minimal understanding of life, a concept of death would not be possible, for inanimate entities would be indistinguishable from dead ones. The reason we don't say of a rock that it's dead, despite the fact that it doesn't exhibit any functionality, is because it's not the sort of thing that one expects to exhibit functionality, that is, it's not dead because *it can't be alive*. For an animal to understand that another is dead, she must classify them first as something one would expect to be alive.

We don't need to understand this minimal notion of life in intellectualistic terms. All we need is for the animal to have some capacity to distinguish animate from inanimate entities (being able to process, for instance, that only the former move in a self-propelled manner and exhibit goal-directed behaviors), as well as for her to be able to generate expectations about the typical functions of beings of one class or another. For instance, to learn that individuals of this particular class move on the ground, are very noisy, and smell of weeds.

This semantic net also comes with the notion of irreversibility, which needs a certain notion of its contrary—reversibility—in order to make any sense. Condition (c) establishes that the animal has an expectation that the dead will not recover these functions. This allows the animal to distinguish dead individuals from those who are asleep, who also lack the (majority of) functions of living beings of their kind, but in this case there is still the expectation that they will exhibit these functions sooner or later.

Third, the minimal concept of death leaves room for variation across cultures, species, and individuals. We have seen that this concept can accommodate the worldview of a Catholic grandmother, an anarcho-punk teenager, and someone who believes in reincarnation, while leaving out people, like Pelayo, who are fundamentally confused about death. In addition, the minimal concept of death allows different species, different individuals, and even the same individual in different periods of her life to have slightly different concepts of death that correspond to how they understand the functionality of living beings of this or that kind.

Fourth, the minimal concept of death allows the animal to make inferences with it. By this I mean that an animal that classifies another as dead will have certain expectations about the individual in question. More specifically, the expectation that she will not exhibit the functions characteristic of living beings of her kind. But if the "corpse" suddenly started to move, the animal would first feel surprised, and then would conclude that the animal in question is not dead. (And perhaps she might even learn to better discriminate dead individuals in the future.)[22] In this respect, the animal is different from our friends the ants, who relentlessly continue with their necrophoresis even when the conspecific they are carrying is displaying signs of life (and perhaps of anger).

Fifth, the animal with a minimal concept of death is also distinguished from ants in that her concept is not tied to any specific sensory stimulus. The minimal concept of death emerges from a more or less general, holistic understanding of what the living usually do, and therefore is not done by the dead. For this reason, it's not triggered by any particular sensory stimulus. In fact, it may be linked to many different ones, as we shall see.

Last, the animal with a minimal concept of death will also differ from ants in that we can't expect her to exhibit a fixed behavioral response to death. The minimal concept of death will be involved in cognitive reactions to death. These, as we saw in the second chapter, are characterized by considerable variability, as well as an interaction with the development and learning history of the individual. Contrary to the case of the ants, we won't be able to know ahead of time how an animal with a minimal concept of death will react when faced with this phenomenon. The type of reaction she exhibits will depend on many other factors, such as her personality, her relationship to the dead individual, her other beliefs and desires in that moment, and so on.

And this is all—for now—with respect to the minimal concept of death. I hope you're still there. With this possibly harsh, though necessary, journey through philosophical analysis, we have reduced the concept of death to a series of indispensable conditions, making sure that it counts as a concept and, at the same time, leaving out all that isn't strictly necessary. This way, we can keep intellectual anthropocentrism at bay.

Having delineated the minimal concept of death, we will be able to answer the question of how extended in nature we can expect it to be. This is because we now have a clear definition that can guide us in this pursuit. We now know that, in order to answer the question of whether Lucy and Bronte knew that that baby was dead, we need to determine whether these two chimpanzees knew that the infant was not displaying the typical functions of individuals of her kind and that her state was irreversible. In the remainder of this book, I will try to convince you that it's more than likely that Lucy and Bronte did indeed have a minimal concept of death and knew that they were interacting with a corpse. And further, I will try to demonstrate that animals of significantly inferior cognitive complexity to chimpanzees' will probably also be capable of acquiring a minimal concept of death. Moreover, I will tackle the question of how complex animals' concept of death can be, that is, whether it can incorporate other subcomponents apart from *nonfunctionality* and *irreversibility*. But first—and please forgive all these preliminaries—we must deal with another form of anthropocentrism that also plagues comparative thanatology: emotional anthropocentrism. To this I devote the following chapter.

5

The Dog Who Mistook His Human for a Snack

In 1997, a thirty-one-year-old German man committed suicide by shooting himself in the head. When his mother and sister found the body, they saw that it was missing parts of its face and neck, which showed signs of having been bitten. The individual who was guilty of this mutilation was none other than the German shepherd with whom the man lived, who was found chilling next to the body with a huge chunk of his human in his stomach.[1]

The behavior of this anonymous dog, whom we can nickname Firuláis, is much more common than we would like to think: a big proportion of humans who die in the sole company of their pets end up as a little something to eat.[2] Perhaps the temptation is to think that hunger is what moves these animals to resort to the only source of nutrition they can find in that moment: their tender and juicy human caretaker. But this couldn't have been Firuláis's motivation. When the body was found, a mere forty-five minutes had passed since the time of death, and his food bowl was still half-full. In fact, around a quarter of all pets who snack on their caregivers do so in the twenty-four hours after the passing and with the animal having access to other food sources.[3]

Instead we may be tempted to think that Firuláis's human companion mistreated him while he was still alive and that there was no affective bond preventing him from being eaten. Moreover, we might assume, this would not happen to us because we are exemplary caretakers. Our dogs would never do something like that; they love us too much. And yet, there's no indication that Firuláis had been mistreated, and in the vast majority of cases there's no history of animal abuse that can explain this behavior.

Is it then possible that our dogs don't actually love us? That they simply see us as a walking piece of meat that they're hoping will croak soon so that they can finally take a bite? For now, the science doesn't support this hypothesis either. On the contrary, everything seems to point to the idea that our dogs really do love us. We know, for instance, that they prefer our company to that of other dogs, that they avoid and don't accept food from humans that they have witnessed treating us badly, and that they're even capable of rescuing us if we're trapped and in need of help.[4]

How then can we explain Firuláis and company's behavior? An explanation may come from the form that this behavior takes. Dogs, just like wolves, are scavengers in addition to hunters. (Although we tend to think of dogs as companion animals, in reality approximately 75 percent of the planet's dogs live on the street without being cared for by any human.)[5] When a dog or a wolf feeds on a carcass she has come across, she usually starts to eat from the abdomen, which is where nutrient-rich organs are to be found, and then follows by feeding on the limbs. Only 10 percent of these cases involve bites to the corpse's face. This contrasts with instances of companion animals who feed on their caregivers. Here, almost three quarters of cases appear with bites on the face, and only 15 percent have wounds on the abdomen.[6]

This fixation on the face suggests that Firuláis's initial motivation was probably not to eat his human, but rather that this behavior started as an attempt to make him react. Our face is the part of our bodies that our canine friends pay the most attention to, for it is key to understanding our emotions and communicating with us.[7] Consequently, it is to be expected that Firuláis, upon seeing his caretaker lying still after the gunshot, began to try to get a reaction from him by nudging his face with his snout. In the absence of a response, and in order to calm himself down or out of sheer frustration, he might have started licking, then nibbling, and once blood was drawn the temptation to take a bite might have been overwhelming.[8] That is, it's likely that Firuláis's love for his keeper and his anguish upon his lack of response were at the root of his behavior.

Despite how intriguing these cases are and the number of them that have been documented, comparative thanatology has not yet picked them up as attention-worthy behaviors. The articles that report on them appear in forensic science journals and not in animal behavior ones. How is this possible, if comparative thanatology aims to study animals' reactions to death?

Emotional Anthropocentrism

Part of the reason why comparative thanatology does not pay attention to behaviors like Firuláis's has to do with *emotional anthropocentrism*. As we saw in the third chapter, emotional anthropocentrism consists of presupposing that the only way of emotionally reacting to death is the human way. In fact, this is not completely precise. It's obvious that thanatologists are aware that there can be many ways of emotionally reacting to death and that they don't have to resemble our own. But what is usually presupposed is that the only way of emotionally reacting to death *that deserves our attention* is the human way. This is emotional anthropocentrism.

This anthropocentric bias leads comparative thanatologists to look for animal reactions to death that resemble human ones. The best example of this comes from one of the foundational texts of this discipline. This paper, which was published in 2010 by James Anderson, Alasdair Gillies, and Louise Lock, documents a case that was observed in a safari park in the UK. An elderly chimpanzee called Pansy died after an illness that had lasted several days.[9] What is interesting for the topic at hand is the fact that the authors explicitly connected the behaviors observed in the remaining chimpanzees to typical human practices surrounding death.

Thus, they pointed out that, during Pansy's final days, the chimpanzees behaved in a respectful and calm manner toward her and adapted her resting place to make her more comfortable, something that the authors relate to the anticipatory grief, respect, and care that humans tend to show to someone who is dying. When Pansy died and the other chimpanzees inspected her mouth and manipulated her limbs, the authors took it as something similar to our practice of testing to see whether someone is displaying signs of breathing or a heartbeat. The aggressive behavior exhibited by one of the chimpanzees, who attacked the corpse, was taken as an attempt to resuscitate Pansy or as an expression of frustration similar to our feelings of anger or denial toward the person who has passed away. The actions of Pansy's daughter, who remained by her side throughout the night, were interpreted as a nighttime vigil, and those of a chimpanzee who spent more time than usual grooming one of her mates were seen as an example of consolation or social support. The apes slept restlessly that night, frequently changing posture, which the authors related to the disturbed sleep of someone who has lost a loved one. The care with which they removed straw from Pansy's body in the morning was interpreted as them cleaning the carcass. And, last, the apathy they showed in the following weeks and the fact that they didn't eat much were seen as a manifestation of the grief that they were experiencing, and their refusal to sleep on the platform where Pansy had died was interpreted as something akin to the human practice of avoiding touching places or objects associated with the one who died.

The authors of this study took these analogies between the behavior of chimpanzees and humans as a reason to set up the discipline of comparative thanatology:

> These behaviours highlight the interest of a comparative evolutionary perspective on death and dying in species without symbolic representations of death or death-related rituals. [...] We propose that chimpanzees' awareness of death has been underestimated [...]. Although data are likely to accumulate slowly, a thanatology of *Pan* [the genus to which chimpanzees and bonobos belong] appears both viable and valuable.[10]

The emotional anthropocentrism I was talking about is perfectly captured by this quote. The authors are saying that what makes these behaviors interesting, and their study valuable, is precisely the fact that they resemble human practices. That is, chimpanzees' reactions to death are not interesting in themselves. They *become* interesting when we start to realize that they are similar to human reactions. If this isn't an anthropocentric viewpoint, I don't know what is.

Emotional anthropocentrism leads thanatologists to look for animal reactions to death that resemble humans'. This has various consequences. To begin with, it leads to an excessive fixation on primates. As you may have noted and will continue to note, many of the empirical examples that I refer to in this book are on primates. This relates to this anthropocentric bias. Most comparative thanatology papers document the behavior of apes and monkeys, and even the very discipline was originally proposed as a thanatology of the genus *Pan* (not just primates, but chimpanzees and bonobos—the ones that are most closely related to us). Even though some interest in the thanatological behaviors of other species is starting to emerge, primates still dominate the debate (literally and figuratively).

Emotional anthropocentrism leads scholars to focus on apes and monkeys because they are the animals that resemble us the most. In fact, even though we sometimes forget about this, we are also primates. We, together with chimpanzees, bonobos, orangutans, and gorillas, are the only great ape species that still exist. Nonhuman primates resemble us anatomically: they have two arms and two legs, hands with five fingers and feet with five toes, and faces that look like our own (nothing like seeing a chimpanzee with alopecia for this resemblance to become evident—as well as for one to realize that one should never attempt to fight these apes; see figure 5). Their gestures and how they relate with each other also make them similar to us. If what we're looking for are reactions to death that remind us of our own, primates, and especially great apes, are the perfect place to go looking for them.

Primates have also received a lot of attention due to intellectual anthropocentrism. If you recall, this is a bias that leads thanatologists to overintellectualize the concept of death. Apes and monkeys are not just our closest living relatives—we can also

FIGURE 5. A chimpanzee with alopecia.

count them among the most intelligent species of the animal kingdom, and more specifically among those with an intelligence that most closely resembles our own. If we presuppose that the concept of death is something that requires human-like cognitive complexity, it's logical for us to seek it among primates.

Emotional anthropocentrism, however, doesn't just lead thanatologists to focus on the study of primates, it also gives rise to an excessive emphasis on affiliative behaviors toward the dead. Affiliative behaviors are those whose function is to establish and strengthen affective and social bonds. They are those practices that express care, affection, or friendship and may be behaviors such as playing, kissing, hugging, grooming, sleeping next to one another, and so on. These types of behaviors, as well as others that are related to the expression of an affective bond, have been considered of interest for comparative thanatology when they are directed toward an individual who is dying or already dead. The reason behind this interest has to do with the search for expressions of *grief*.

Animal Grief

Grief is an emotional process that humans go through when someone close to them passes away. Although we tend to think of it as an emotion, in fact it is a collection of emotions, for it is a process that extends in time and changes shape, incorporating different feelings such as despair, anger, gratitude, love, sadness, or confusion.[11] However, when we think of grief we usually have in mind an emotion that is primarily of sadness, an intense sorrow caused by the death of a loved one. We consider this sadness as the death-related emotion par excellence, and this is what comparative thanatologists who are in the grip of emotional anthropocentrism look for in animals.

Even though it has its origins in emotional anthropocentrism, there's nothing intrinsically bad about the scientific project of trying to find evidence of grief in other species. The problem, as we shall see, comes from an *excessive* focus on this emotional reaction to death. But the search for grief in animals is an interesting project in and of itself. In fact, everything seems to point to the idea that this phenomenon is present beyond the *sapiens* species.[12]

One of the first anecdotes on animal grief was collected by Jane Goodall during her observations of the chimpanzees of Gombe. There she met a chimpanzee that she named Flo and her son, Flint, both of whom were very close. When Flo died, Goodall witnessed what appeared to be a case of grief:

> Flint [...] was eight and a half when old Flo died, and should have been able to look after himself. But, dependent as he was on his mother, it seemed that he had no will to survive without her. [...] Flint became increasingly lethargic, refused most food and, with his immune system thus weakened, fell sick. The last time I saw him alive, he was hollow-eyed, gaunt and utterly depressed, huddled in the vegetation close to where Flo had died. Of course, we tried to help him. I had to leave Gombe soon after Flo's death, but one or other of the students or field assistants stayed with Flint each day, keeping him company, tempting him with all kinds of foods. But nothing made up for the loss of Flo. The last short journey he made, pausing to rest every few feet, was to the very place where Flo's body had lain. There he stayed for several hours, sometimes staring into the water. He struggled on a little further, then curled up—and never moved again.[13]

Although this is an especially dramatic example, in more recent literature we can find many more anecdotes that suggest that nonhuman primates are capable of experiencing grief.

One example comes from Amy Porter and her collaborators.[14] In this paper they documented the death of various gorillas of the group they were monitoring. One of them was Tuck, a female who was thirty-eight years old at the time of death. Her son, Segasira, a five-year-old gorilla, displayed a response that may have been an expression of grief. Tuck and Segasira had a very close relationship and, when she died, he

FIGURE 6. Segasira grooming his mother's corpse.

was the member of the group that stayed next to the body for the longest time. The researchers saw him sitting beside his mother's corpse, sleeping next to it, sitting on it, hugging it, grooming it, closely inspecting its face, carefully moving its head (see figure 6). They also witnessed Segasira trying to suckle from Tuck's breast, despite the fact that he had been weaned some time ago. This last behavior is especially noteworthy. Breastfeeding is associated with the production of oxytocin, a hormone that helps us feel better by regulating our stress levels. The researchers interpreted this action by Segasira as a search for consolation; an attempt to calm his own anguish upon his mother's lack of response.[15]

Another suggestive example was noted by Bin Yang, James Anderson, and Bao-Guo Li, who documented the death of a female snub-nosed monkey.[16] The monkey, nicknamed DM, fell from a tree and it took her an hour and a half to die, during which time she was mostly lying on the ground, occasionally twitching and groaning faintly. As soon as she fell, the rest of the members of the group immediately descended from the trees and began alarm-calling. They spent fifty minutes caring for DM, observing and sniffing her face, grooming her, embracing her, gently pulling on her hands, occasionally sending out alarm calls, and preventing the youngest members of the group from getting too close. After a while, they lost interest and left the body's side, all except a male nicknamed ZBD, who had a very strong bond with DM. Together with some other females from the group, ZBD remained sitting next to the body, but in contrast to the former, who were busy grooming each other, he was paying attention to DM's body, gently touching and grooming it. After a while, the group decided to leave the area, and DM attempted to get up but finally collapsed and died. ZBD remained next to the body for several minutes, stroking it and pulling on one of its hands. Then he began to walk in the direction of the group, turning back from time to time to look at DM's body. After this, he sat and spent a few minutes alternating his gaze between DM and the rest of the group until, finally, he left the corpse behind and reunited with the others.

We also see instances of apparent grief beyond primates, for instance, in elephants. Iain Douglas-Hamilton and coauthors witnessed such a case with the death of one group's matriarch, Eleanor.[17] This African elephant collapsed one day and died on the following one due to what seemed to be natural causes. Many elephants of different families showed an interest in her while she was dying, as well as once she was already dead. This may be due to elephants' natural death-related curiosity (I will discuss this in the following chapter) or to their well-known tendency toward empathy and compassion.

For instance, Grace, a matriarch from another family, came across Eleanor when she had just collapsed and tried by all

FIGURE 7. Grace attempting to lift Eleanor up.

means to help her get up and stand upright, appearing visibly agitated the whole time (see figure 7). Once Eleanor had died, many elephants were observed reappearing time and time again at the spot where her body lay. Among them was the elephant Maya, who had spent the most time next to Eleanor while she was still alive and whom the authors suspected to be her daughter. Although they didn't see Maya interacting with the body, they did see her coming back to its surroundings day after day, and spending more time in close proximity to the corpse than any other elephant, even when it was already being taken apart by scavengers. Although she wasn't seen performing any activity while she was in the location of the body, it can be assumed that the strong affective bond that she felt for Eleanor was what kept her coming again and again to the same place.

But what about animals who are not as famously smart as primates and elephants? We can find indications of grief here as well. Zoe Muller, for instance, reported the case of some giraffes reacting to a dead calf.[18] The little giraffe had been born with a deformed leg and her mother had always remained within twenty meters of her, offering her a degree of attention and care that is uncommon in this species. However, when the calf was but a few weeks old, she died of natural causes. When Muller came across the corpse, it was surrounded by sixteen giraffes, all of whom were female and appeared very distressed, circling the body, inspecting it, sniffing it, nudging it with their snouts. Among them was the baby's mother. When Muller returned on the following day, the troop of giraffes was still circling the carcass, which remained intact thanks to this level of protection. On the third day, Muller came back and could not see any giraffes in the area. After scanning the surroundings with her binoculars for a while, she finally saw the mother, who wasn't at the same spot but some fifty meters away. When Muller managed to get closer, she saw that the calf's remains were now at this location and had been mostly devoured by scavengers. When Muller returned on the following day, the corpse was nowhere to be seen, but the mother was still wandering about the area.

Another suggestive example comes from Dante de Kort and colleagues, and refers this time to a group of collared peccaries, animals that resemble boars and who live in woods throughout North, Central, and South America.[19] The researchers came across a dead female peccary, and saw that two members of her herd were sleeping next to her and another two were in the surroundings. Because this appeared to be an unusual behavior, they decided to set up a camera trap to monitor the peccaries' actions in the days that followed. What they found was that, despite the fact that during the days the peccaries went about their business as usual, they always returned to the location of the corpse, especially at night. The researchers, who didn't know if the peccaries were related, watched them pushing the carcass, nudging it with their snout, observing it, sniffing it, biting it, grooming it, and attempting to make it stand upright. It was also very common to see them sleeping in close proximity to the corpse, often snuggling next to it. After ten days, a pack of coyotes appeared and attempted to snatch the delicacy. The peccaries managed to shoo them away a few times, but during the night the coyotes finally managed to attack the corpse and started eating it. The remaining peccaries fled and were not seen in the area again.

As I said, there's nothing wrong with debating whether animals are capable of experiencing grief. What's crucial, however, is not to lose sight of what grief can tell us about the cognitive and emotional capacities of animals. As is evident from these examples, when we see a case of apparent grief, the animal who exhibits it tends to have a very strong affective bond with the deceased. And this isn't coincidental. In the words of anthropologist Barbara King, who has extensively studied this phenomenon: "Where we find animal grief, we are likely to find animal love, and vice versa. It's as if the two share emotional borders."[20] Grief, understood as a reaction of distress or sadness upon another's death, presupposes that we *care* about the other, that there is an emotional bond that connects us to them.

Grief, therefore, is something relevant for us to focus on if what we're looking for is evidence of animals' capacity to feel love, care, affection, or friendship. But what if what we're looking for is evidence of a concept of death? In this case, it's not as clear that concentrating on grief is going to help us much.

In principle, grief may appear in the absence of a concept of death, because it may be caused by the mere *absence* of a loved one. King tells us the story of Willa, the cat of some friends of hers who appeared to go through a horrible period of grief after the death of her sister, Carson, with whom she'd lived for fourteen years.[21] When Carson died, Willa began to act in a very strange manner. She roamed their house all day, ceaselessly looking for her absent sister and making some noises that her keepers had never heard before and which they described as keening. It's perhaps legitimate to talk in this case of grief, but we can't speak of a concept of death, for Carson died in the veterinary clinic and Willa never saw or interacted with the corpse, and so she had no way of knowing that she was dead (and for reasons we shall see in the following chapter, it's not terribly likely that Willa had a concept of death, even a minimal one).

However, the other cases that we have seen in this chapter happened in nature. In all of these examples, the animals exhibiting the apparent mourning behavior did have the chance to see and interact with the corpse, so in principle there might have been a concept of death involved.

Even so, it's still problematic to focus on cases of grief if we're looking for evidence of a concept of death in nature. This relates to the fact that grief may lead animals to interact with the deceased *as though* they were alive. We can illustrate this particularly well with reference to what's known as "deceased-infant carrying," one of the behaviors that has received the most attention in the thanatological literature.

Deceased-Infant Carrying

The practice of holding on to a baby's corpse for prolonged periods of time has been most extensively documented in primates.[22] In this case, the fact that almost all the evidence comes from primates probably doesn't have as much to do with emotional anthropocentrism as with the fact that apes and monkeys have hands, which makes it significantly easier to go through life with a carcass on you. Still, there have been a few cases witnessed in nonprimate species. For instance, Rob Appleby, Bradley Smith, and Darryl Jones observed a dingo mother carrying the corpse of one of her pups in her mouth for three days while she moved from one place to another with her litter.[23] And in the case of cetaceans, there are many examples of mothers, like Tahlequah, who carry their dead calves, even when these are in an advanced state of decomposition. Mothers often stop feeding to focus all their attention on the carcass and protect it from their conspecifics and from potential scavengers while they carry it with them for kilometers.[24]

Deceased-infant carrying in primates has received an immense amount of attention in the literature, due to the fact that it's a very striking behavior and that its occurrence seems hard to explain. We can illustrate this with a particular example, which was reported by Arianna de Marco, Roberto Cozzolino, and Bernard Thierry.[25] This is the case of a Tonkean macaque called Evalyne who is part of a captive population in an Italian zoo. Before this particular case took place, ten other females of the group had been witnessed losing their infants and carrying their corpses for periods that ranged from a few hours to seven days. Evalyne's case, however, was particularly surprising, for her behavior extended for a whopping twenty-five days. What follows is a reconstruction of the sequence of events, as narrated by de Marco and collaborators (see also figure 8).

The infant was born seemingly healthy and received normal care from Evalyne, who was a first-time mother. However, five days after the birth, Evalyne woke up to find her baby had died. On the following morning, Evalyne did not attend the daily food distribution, and her caretakers noticed that she was very agitated and kept looking intently and screaming at her own

FIGURE 8. The evolution of Evalyne's behavior throughout the first twenty-four days.

reflection in the plastic door that connected the indoor and outdoor enclosures. She had never been witnessed behaving this way. During the week that followed, she appeared much calmer but carried her baby's carcass wherever she went, exhibiting care behavior, grooming and licking it, paying special attention to its face. She was also observed putting her fingers and her tongue inside the corpse's mouth, something that mothers of this species do to try to induce a suckling reflex. Despite the intense smell emanating from the tiny body, Evalyne did not let go. The remaining monkeys paid no attention to the corpse.

As the days went by, the body began to progressively dry up until it reached a state of mummification. Evalyne carried it with her at all times, holding it with one of her hands, with a foot, or in her mouth. If she dropped it, she would immediately grab it back up. On the fourteenth day, the carcass lost its skull, but this did not deter Evalyne. During the week that followed, she continued carrying the remains, which kept losing fragments bit by bit. She was often seen bringing them to her face, perhaps with the intention of sniffing them or inspecting them up close.

On the eighteenth day, Evalyne let go of the carcass for the very first time but didn't stop watching over it. While it was on the ground, an adult female came close to it and briefly smelled it, after which an adolescent female grabbed it and played with Evalyne for a bit while she had it in her hand.

On the nineteenth day, Evalyne was seen for the first time nibbling at the remains of the corpse and eating little bits of it.

On the twenty-second day, the body fell apart and different pieces of it were dispersed throughout the enclosure. Evalyne began to be continually seen with one of the pieces either in her mouth or in her hand, sometimes dropping it briefly on the ground only to pick it up again later. If she came across another of the pieces, she would often exchange it for the one she was carrying in that moment. She was frequently seen gnawing at what remained of the body and eating small fragments of it. On the twenty-fifth day, she was seen for the last time carrying a part of something that might have been a leg or an arm, with no other piece visible in the enclosure. None of the other macaques were ever seen carrying any of the remains.

Evalyne's behavior is especially noteworthy due to its duration and persistent nature, but it is by no means an exception. Deceased-infant carrying has been widely documented in primate mothers who have lost their babies. However, as we saw in the case of Bronte, it's not always as obsessive a behavior as Evalyne's. Some mothers carry their dead offspring for just a few hours or days, while others don't engage in this behavior at all, and still others prolong it for months.

Deceased-infant carrying is a behavior that has received tons of attention due to the apparent difficulty of explaining it. In a recent paper, Claire Watson and Tetsuro Matsuzawa tried to find a single hypothesis that could explain all the reported cases, and they argued that this isn't possible.[26] According to these authors, the most commonly accepted hypothesis, namely the idea that this is a behavior that is triggered by the mother-infant bond, would not explain cases, such as Evalyne's, that end up in cannibalism, and much less those in which cannibalism is alternated with caregiving behavior. The mother-infant bond hypothesis would also not be able to explain those cases in which females other than the mother, or even males, have been seen carrying infant corpses.

It has also been postulated that perhaps first-time mothers or those who lose a baby that they were strongly bonded to would be more likely to carry out this behavior. And yet, there doesn't seem to be a reliable correlation between these factors.

Other authors have postulated that climate may be a determining factor, for some conditions allow the mummification of the carcass, which, as in the case of Evalyne's baby, allows it to maintain its shape and thus makes it more recognizable as an infant, as well as facilitating carrying. Against this hypothesis, Watson and Matsuzawa point out that many cases of deceased-infant carrying in humid and tropical climates have been documented, and there are also reports of mothers carrying corpses that are deformed, bloated due to decomposition, or of which only an unrecognizable piece remains.

Yet another hypothesis postulates that it might be adaptive for the mothers to hold on to the body just in case it regains consciousness. Given the energy already invested in the baby, it might be a good idea to adopt a "wait and see" strategy. However, as Watson and Matsuzawa also point out, there are many additional reasons why this behavior is maladaptive: a decomposing corpse is a source of pathogens, keeping one hand busy is risky for species who live on trees, carrying a carcass with you at all times is energetically demanding, and obsessively caring for a dead baby may impede copulation or resumption of the menstrual cycle. All of these factors can ultimately prevent the mother from passing on her genes, thus resulting in a maladaptive behavior.

Although it's certainly interesting that one can't find a fixed rule that explains this behavior or allows us to predict when, how, and on the part of whom it's going to take place, the truth is that it's simply not realistic to expect such a finding, given that apes and monkeys—as well as the other species in which deceased-infant carrying has been documented—are animals of extreme cognitive and emotional complexity, with their own personalities and life histories that determine how they will act in a given situation. Returning to the distinction I introduced in the second chapter, what we see in these species are cognitive rather than purely stereotypical reactions to death, so it's to be expected that there will be high variability in how these behaviors happen, as well as for us not to be able to find a single explanation that encompasses all cases.

However, what we *can* say is that the *majority* of cases occur in mothers who have lost their babies, and that the *main cause* of this behavior finds its roots in the hormonal and emotional mechanisms that are responsible for the mother-infant bond. This doesn't mean that there can't be other factors, both circumstantial and personal in nature (such as the climate, the mother's past experiences, the cause of death, etc.) that may influence how this behavior manifests itself, or even whether it manifests at all.

The fact that the mother-infant bond is involved in deceased-infant carrying is evident for various reasons. The first one, which may seem trivial, is that this behavior has to be triggered by mechanisms present in the mother, for the infant, being dead, can't be "doing" anything that motivates the behavior.[27] At the same time, unless the corpse is mangled or disfigured, its very appearance or smell may be enough for these mechanisms to be triggered.

We also know that the motivation behind this behavior must be very intense. Nonhuman primates are not completely bipedal, and a dead infant, for obvious reasons, doesn't cling to the mother like a live one would, so carrying the body requires considerable effort. In addition, as I already mentioned, it's dangerous to have one hand permanently occupied, especially in the case of arboreal species, which applies to many primates. In cetaceans, the motivation may be even stronger, for due to their lack of hands this becomes an even more difficult behavior to carry out for prolonged periods of time.

Mothers who carry their dead infants also tend to fiercely protest any attempt at a forced separation from the carcass, something that powerfully reminds us of their reactions to any threat to their babies' safety. Watson and Matsuzawa tell us of the case of a captive macaque who, after carrying her dead infant for twenty-two days, began to leave it on the ground while she was doing her things, always keeping within twenty meters of it. Assuming that she had lost interest in the corpse, the zookeepers attempted to remove it from the enclosure. When they put it back in after ten minutes, the monkey immediately ran up to the body and fiercely snatched it back while she carried out threat displays toward the humans who had stolen it from her.[28]

Similarly, Melissa Reggente and colleagues narrate the case of a female bottlenose dolphin who was seen carrying a juvenile corpse in an advanced state of decomposition. The scientists who witnessed it decided to tie the body up with a rope and take it back to shore in order to bury it. (One can't help but be reminded of ants and their necrophoresis when imagining these researchers engaged in such an absurd task.) The mother decided to swim behind the boat, circling the corpse and insistently touching it with her beak until the waters were too shallow for her to continue. Hours after the carcass had been buried, the dolphin was still obstinately swimming around the area.[29]

The intensity of these mothers' motivation suggests that there are very primary and very powerful mechanisms at the base of their behavior, and that these probably coincide with the mechanisms that are responsible for the care that the females of these species display toward their young. However, this same intense motivation, coupled with the caregiving behaviors that these mothers direct toward the corpses, has made some authors ask themselves whether the mothers actually know that their baby is dead, arguing that perhaps they carry the body so relentlessly and continue to nurture it because they believe it to be alive.[30] Is this a plausible hypothesis? This is where things get tricky, and where it starts to become clear how problematic it is to focus on mourning if what we're looking for is a concept of death in nature.

We definitely do have some indications that the mothers know that something is up with their infant. Although we do see some caregiving behaviors directed at the corpses, we also see carrying techniques that are not adopted with live infants. Mothers often carry the carcasses upside down, in their mouths, or hanging by a limb and dragged across the ground (see figure 9). All of this may stem from an absence of "cooperation" on behalf of the baby, who would normally grab on tight to her mother, but it may also be evidence that the mothers have registered the death of the little one.

Although in the next chapter I will present a few reasons to think that these mothers do in fact have a concept of death and know what is going on with their infants, the truth is that this behavior is not a good place to look for evidence of a concept of death. This is because the affective bond that unites a mother with her baby may lead her to want to treat her *as if* she were still alive.

In fact, this is a behavior to which *human* mothers are drawn and that is even recommended as a therapy to cope with the loss of a baby. In a heartbreaking article, Ana Todorović describes her experience after the birth of her second daughter, who was stillborn. She tells us that for several days she couldn't help treating Nadia, her baby, as though she were alive, bathing her, dressing her, stroking her, sleeping next to her—a behavior that the medical team allowed and even encouraged, for it's

FIGURE 9. Chimpanzee mothers carrying their dead infants with techniques not used with live ones.

often considered to be therapeutic.[31] In Todorović's harrowing description of her feelings we can perhaps see a glimpse of something similar to what these nonhuman mothers may be experiencing when they carry, groom, nurture, and protect their babies during days, weeks, or months:

> I'd like to think that I'm a rational person, but my devastation was punctuated with brief flashes of hope as I watched her, over the next few days, when there was clearly no doubt about her state. Someone would walk near where she lay in her cot, and I would instantly twitch towards her, watching

for any sign of motion. [...] I knew, yet I couldn't help hoping. It was a deep-seated, visceral response to her presence. [...] I kept wanting not to see her anymore, and wanting to see her again, for it to be over and for it to never end. I liked holding her little hands. I liked stroking her feet. But mostly, I liked touching her face. Even when it became cold, the skin on her cheeks was as soft as only a newborn's can be. [...] The act of giving birth was a strong force in consolidating my emotions towards Nadia. [...] Perhaps this love is just there, simply felt. Unconditional, even on knowing her. It's not a hope, it's not a thought. Perhaps its building blocks are too far removed from any explanations, are too deep in our roots, go further back in time than we can meaningfully contemplate on. My mind will try to build some meaning around her because my feelings take me in that direction, but the idea of a child that *almost was* is something too elusive to be carefully taken apart and then mentally rebuilt. It's just a gaping hole, a hollow, aching absence [...].[32]

The apparently irrational behavior that we see in Todorović, in Evalyne, in Bronte, or in Tahlequah responds to the type of reproductive strategy followed by the species to which these individuals belong. Both primates and cetaceans are part of what scientists call K-strategists, which are those animals that have few offspring and devote huge levels of care to each one of them in order to increase their chances of reaching maturity. This contrasts with the approach followed by r-strategists, which is the group to which most animal species belong.* These animals don't offer any care at all to their offspring, but

---

* The letters K and r refer to two constants in a population equation that you don't need to be familiar with.

compensate this by having many of them at a time, thereby ensuring that at least one or two of them will survive.

In *K*-strategists, it's crucial that there be some extremely powerful mechanisms governing maternal care, for without them the young would have very low chances of reaching adulthood. (Think of how long a human baby would survive if left alone in the woods.) The solution that evolution has come up with is precisely this impossibly strong, unconditional, and irrational maternal love. This is why deceased-infant carrying is a bad place to look when searching for evidence of a concept of death, for this behavior seems to be governed by irrational mechanisms, and it's difficult to predict beforehand how it will interact with a concept of death. Perhaps knowing that their baby is dead gives these mothers a reason to abandon the body, but perhaps, as in the case of Todorović, it gives them a reason to continue seeking out that contact.

But what about the cannibalism that we see in some of these mothers? Here our emotional anthropocentrism blinds us and, like in the case of Firuláis, makes it very difficult to conceive how that behavior can be made compatible with an affective bond. But the truth is that, even though in our societies it's absolutely taboo and inconceivable to eat our offspring, in principle it's possible for the mother's love for her baby to be what's motivating this behavior. Don't we sometimes feel a certain primary impulse to "eat" those we love the most?* Perhaps in these animals this impulse manifests itself to its ultimate consequences. Or maybe this behavior has something to do with seeking out consolation. The same way Segasira tried to suckle from his dead mother's breast, maybe these mothers eat their

---

* In Spanish we even have the expression ¡Te como! ("I eat you!"), which we use to express affection.

young in an attempt to calm their own anguish. In fact, we know that carrying the corpse helps them to regulate the stress caused by the death of their baby, which is a reason why it has been postulated that this behavior is a form of "comfort contact."[33] Ultimately, love can manifest in many different ways, and though some may be alien to us, that's not enough of a reason to rule out a grief-based explanation for these behaviors.

Grief and the Concept of Death

As is illustrated by the debate on deceased-infant carrying, the question of whether animals are capable of experiencing grief is an interesting one and one to which we should probably respond in the affirmative. However, it's important not to confuse it with the question of whether animals have a concept of death. This is not only due to the fact that, as we have seen, it can be difficult to distinguish if there's a concept of death when there's grief, but also and primarily due to the theoretical reason that they are *two different questions*.

One is the question of whether animals feel grief, and the other, the question of whether they understand death. Unfortunately, in the thanatological literature these have not been properly treated as independent issues. The question of animal grief is constantly confused with the question of the concept of death in animals. This stems from emotional anthropocentrism. Given that grief is the prototypical human response to death, we expect animals who understand death to exhibit grief, but this need not necessarily be so. In the same way that there can be grief without a concept of death, there can be a concept of death without grief.

To illustrate this, let me tell you a short story. When my brother was a kid, one day he brought home a tiny slug that he had found in the garden. He was exhilarated by his discovery; he held the minute slug in his grubby little hand while he proudly said to my mother: "Look, Mom! It's my new pet, his name is Babosín!"* My mother, horrified by the presence in her home of this creature who looked like a mutant booger, grabbed poor Babosín with a piece of paper and flushed it down the toilet without hesitation, unperturbed by the prospect of her son's imminent tantrum. I still remember my brother's cry as the flush took the mollusc's life: "Noooo! Babosíííín!" The trauma was such that, when choosing his next pet, my brother opted for a plastic lizard instead.

The purpose of my sharing this anecdote is not just to embarrass my brother, but also to illustrate that no concrete emotion needs to follow from a concept of death. For my brother, Babosín's demise was a tragedy. For my mother, it was a relief. And yet, both had correctly processed that Babosín had died.

This idea also applies to nonhuman animals. Grief is not an indicator of a concept of death, but rather a sign that there is a very strong affective bond toward the deceased. But an animal may in principle experience all sorts of emotions upon another's demise. She may be happy, if the death means that she will rise in the social hierarchy. She may feel scared, if she interprets it as a sign that there is some danger nearby. She may feel a purely selfish form of sadness, if it entails the loss of the individual who best scratched her back. She may feel excitement or hunger, if she perceives the corpse as a food source. Or she may even feel indifference, if the one who died meant nothing to her. The concept of death is compatible with a very wide range of emotional reactions, only one of which is grief.

---

* This very sophisticated name literally translates as "Little Slug."

If we focus excessively on grief, we may be potentially wasting many opportunities for uncovering a concept of death in nature. Allowing ourselves to be led by emotional anthropocentrism will also make it harder for us to pay attention to behaviors that are difficult to understand from a human viewpoint, such as cases like Firuláis's or behaviors like cannibalism, infanticide, or necrophilia. And yet, these behaviors are prevalent in nature and may contain important clues to allow us to decipher how animals experience and understand death.

In the next two chapters, we shall see how, once we set aside intellectual and emotional anthropocentrism, it's straightforward that the concept of death will likely be very easy to acquire and thus widely present among nonhuman animals, even though the ways in which they react to death are much more varied than one would expect from a purely anthropocentric standpoint.

6

The Elephant Who Collected Ivory

If, like me, you grew up obsessively watching *The Lion King*, you will likely be familiar with the notion of elephant graveyards: those places where these pachyderms supposedly go to die, which are dark and full of terrors, and sometimes hide packs of hyenas and lions with scars on their faces and malicious intentions.

I regret to inform you that, unfortunately, it seems these graveyards are a myth. Although it's true that occasionally the bones of different elephants have been found in a relatively contained area, this is thought to have been due to intensive hunting or to periods of drought that caused many deaths in a row, and not to these animals having gone to that specific location to expire.[1]

However, it does seem true that elephants—at least African ones—have a singular relationship with death. As we have seen, members of this species are very prone to reacting with empathy and compassion in situations in which a conspecific is distressed or in danger. This extends to circumstances in which an elephant is dying, as we saw in the previous chapter with the case of Grace, who tried to help Eleanor in her last moments despite their being unrelated.

Elephants' interest in death extends beyond compassion. As we also saw with Eleanor's story, elephant corpses tend to receive numerous visits from both related and unrelated conspecifics. However, this could in principle happen because elephants tend to travel through similar routes, so they might come across the corpses just by chance. It is also possible that they're simply attracted to carcasses on account of them being novel objects.

In order to test these hypotheses, Karen McComb, Lucy Baker, and Cynthia Moss carried out an experiment in Amboseli National Park, in Kenya.[2] This experiment consisted of presenting wild elephants with groups of three objects, in order to study their reactions and see which ones generated greater interest in them. Specifically, the experiment had three different conditions, depending on the objects present. In the first of these, they were confronted with an elephant skull, a piece of ivory, and a piece of wood. The elephants showed much more interest in the ivory than in the other two objects, and more in the skull than in the piece of wood. In the second condition, the present objects were the skulls of an elephant, a buffalo, and a rhinoceros. The elephants once again appeared more interested in the skull of their conspecific and showed an equally low interest in the other two skulls. Last, they were presented with three elephant skulls: two of them belonging to matriarchs of other families and one of them belonging to a recently deceased matriarch of their own family. In this case, the subjects appeared equally interested in the three objects.

The results of this experiment demonstrate that elephants' visits to their conspecifics' corpses are not a coincidence. Elephant skulls appear to interest them much more than other natural objects, as well as more than skulls of members of other species. Nevertheless, they don't seem to make distinctions between the skulls of related and unknown elephants. This may be due to the individual not being recognizable merely by means of her skull—though given these animals' powerful sense of smell, they might not need more than bones to identify a concrete elephant—but it also points to the idea that the carcasses of conspecifics are simply interesting to them, which would explain why Eleanor received so many visits even when her death was recent and her corpse, fresh and clearly recognizable.

Something that is especially noteworthy is the apparent fascination that the elephants displayed toward the piece of ivory, despite it being in principle the simplest item. It wasn't just the object to which they paid the most attention, they also manipulated it with their feet much more than any of the others. Beyond this experiment, elephants have been seen transporting the tusks of dead conspecifics, as though they had some sort of obsession with collecting them.[3] One reason why ivory is so attractive to them might be that tusks are a characteristic of elephants that "survives" death. In their social interactions, elephants frequently touch each other's tusks with their trunks, and it wouldn't be surprising for elephants to be able to know, upon encountering a piece of ivory, that it's something that once belonged to an elephant, or perhaps even to this or that particular individual.[4]

Elephants' interactions with the dead appear to go beyond mere curiosity. It's common to see them attempting to lift or even carry conspecifics who are dead or dying. Researchers have also reported on cases of elephants sticking food in the mouth of a conspecific corpse or trying to copulate with it, as well as cases of them watching over and protecting a carcass against predators or other elephants. And, most strangely, elephants have also been witnessed covering carcasses with earth and vegetation, in what resembles a rudimentary burial; a behavior that, in contrast with the others, they have never been seen displaying toward live individuals.[5]

The complexity of elephants' reactions to death is evidence that they are good candidates for possessing a concept of death. As we already saw, variability and complexity in behaviors toward corpses are a sign of the presence of cognitive reactions, which are a prerequisite for the concept of death to be possible. However, there are more reasons for thinking that elephants are well positioned to develop an understanding of mortality.

To see this, it's necessary for us to introduce some notions regarding how a concept of death might emerge in nature. The development of this concept on the part of an animal does not happen by chance, but rather emerges as a result of the interaction of three fundamental causal factors: COGNITION, EXPERIENCE, and EMOTION. We can call this the holy trinity of the concept of death.[6] Let's see what it consists of.

The Holy Trinity of the Concept of Death

The first element in the holy trinity is COGNITION. We need the animal to be capable of having cognitive and not merely stereotypical reactions to death. Cognitive reactions suggest that an animal is capable of possessing concepts. Without the ability to possess concepts, logically the concept of death would be out of reach for any individual. In addition, as we saw in chapter 4, understanding death requires, at the minimal level, being capable of processing the irreversible non-functionality that comes with it. As we shall see in the pages that follow, this doesn't require a huge degree of cognitive complexity, but there is a certain basal level of cognition that an animal must have if she's to be capable of reaching even a minimal concept of death. If an animal is simply *unable* to process non-functionality and irreversibility, then she will never be able to understand death.

The second element is EXPERIENCE. As we already know, some animals, like ants, are born with certain pre-programmed and stereotypical reactions to death. These behaviors are innate and don't require learning. The concept of death is different. Given the way in which we're understanding it here, the concept of death is never going to be innate, but instead it will be the result of direct experiences with death that lead to a *learning* process.

An obvious exception to this are humans, for we have an outstandingly complex communication system with which we can teach and learn about death without the need for direct experiences with it. This is how the majority of children in Western, urban societies learn about death: not because they have had a corpse in front of them and have had the opportunity to poke it with a stick until they reach, by their own means, a concept of irreversible non-functionality. Instead, their learning is usually mediated by storybooks, films, video games, and other cultural manifestations, as well as by the dialogs with their parents that emerge as a result ("What happened to Bambi's mother?" and other awkward conversations).

We will assume here that animals don't have this communicative support in their learning process regarding death. In actual fact, we still have a lot to learn about non-human communicative systems, so there might come a time when science makes this assumption obsolete. The communicative abilities of some cetaceans, for instance, appear to be very complex and perhaps they might sustain teaching about death. However, our knowledge in this field is still in its infancy, so we shall proceed cautiously on the assumption that animals can't communicate with each other about death.[7] As a result, we will presuppose that, in the absence of direct experiences with death, an animal can't acquire this concept.

The third component of the holy trinity is EMOTION. As we saw in the previous chapter, the concept of death is not linked to any concrete emotion. Nevertheless, this doesn't mean that emotions aren't involved in learning about death. Nothing could be further from the truth. Emotions are crucial because they serve to modulate one's *attention*.[8] It's not enough for an animal to have experiences with death—we need her to pay attention to these experiences in order to learn about non-functionality and irreversibility. This is especially important in the case of irreversibility, for it's not a notion that an animal can instantly acquire just by quickly scanning a corpse, but rather it requires time, so it is essential that the issue grabs her attention. For this, some emotion, or affective state more generally, is needed to act as an intermediary and ensure this interest. However, the quality of the feeling is in principle open: love or grief might do, but also curiosity, confusion, excitement, hunger, fear, frustration, etc.

The holy trinity of the concept of death is of course secular, but it's holy and triune because COGNITION, EXPERIENCE, and EMOTION must all be present for an animal to develop a concept of death. However, these three causal factors interact and influence each other in such a way that the high presence of one can compensate for the relative absence of the other two. Thus, for instance, an animal who was especially intelligent might not need too much experience with death to learn about it, and an animal who's had tons of experience with death might develop a minimal concept even if she's not particularly bright. The same happens with emotion: those deaths that generate little interest might still provide the optimal learning conditions if this low interest is compensated by high degrees of experience and cognition. And, conversely, an animal, such as Segasira or Flint, who loses someone to whom she was extremely bonded might be so motivated to stay close to the carcass and pay attention to it that she might start to learn about death with that single experience.

Coming back to elephants, they are good candidates for a concept of death because we see in them high levels of COGNITION, EXPERIENCE, and EMOTION. As their popular image suggests, these are exceptionally intelligent animals with outstanding memories. The relation between the size of their brains and their bodies is comparable to that of great apes, and they display very impressive numerical cognition, spatial navigation, individual recognition, communication, and cooperation, to name just some examples of their abilities.[9]

In the wild, it's to be expected that elephants will have ample opportunities to learn about death, given that they are very long-lived animals, with a life expectancy of around seventy-five years.[10] This, in addition to their powerful memory, makes it more than likely that they will accumulate the necessary experiences to learn about death.

In addition, we have to consider the emotional factor. Elephants, as we have seen, are capable of empathizing with one another, which may motivate them to pay attention to individuals who are dying or already dead. Moreover, these pachyderms are K-strategists, which, as you will recall, means that they have few offspring but devote huge amounts of care to each one. They have one of the longest pregnancies of the animal kingdom (twenty-two months), which adds to a relatively slow development.[11] This means that each calf amounts to an immense resource investment, which gives rise to equally intense levels of care that are shared among the different females of the family unit. If we add to this the fact that a quarter of all calves don't make it past their first year of life,[12] this creates the optimal circumstances for these animals to reach a concept of death with relative ease. Due to this extremely strong emotional bond toward their young, they will be highly motivated to pay attention when an unfortunate demise takes place.

It may be tempting to think that another reason why elephants are good candidates for the possession of a concept of death is the fact that they are *social* animals. Indeed, sociality has been postulated by various thanatologists as a key factor in the emergence of a concept of death in nature.[13] Giovanni Bearzi and coauthors, for instance, argue that with sociality come strong emotional bonds, and that it can't be expected for these to simply disappear upon the death of one of the individuals. These emotional bonds, in turn, give animals a reason to pay attention to the corpses of those who were close to them in life, thus triggering a learning process about death.[14] Something similar is defended by Fred Bercovitch, who points out that the cases that are reported in the thanatological literature refer almost exclusively to animals who live in complex social groups, and that no published papers discuss the reactions to death of solitary animals, like koalas or moose, or animals who live in giant herds, like wildebeest or caribou.[15]

It has also been argued that the fission-fusion dynamics that characterize some animal species could be crucial in the development of a concept of death. These dynamics, which happen when animal societies regularly divide up into temporal subgroups, only to later reunite, require the animals to be constantly monitoring and updating the status of other members of the group.[16] They are fluid and flexible societies that need individuals to continually renegotiate their relationships with others. This could incentivize their paying attention to the deaths of their mates, for they can lead to an important shift in one's own and others' status within the group.

Sociality is without a doubt a facilitating factor in some cases. However, it's not crucial enough to justify us turning the holy trinity into a holy square. In fact, sociality is just a reliable indicator of a species in which the concept of death is very likely to emerge insofar as it correlates with high degrees of COGNITION, EMOTION, and EXPERIENCE, which it often does. Let's look at this.

Sociality often correlates with high COGNITION. This is first due to the fact that complex social systems appear to have been an important driving factor in the evolution of certain kinds of intelligence.[17] Living in a complex social group entails a series of challenges that are related to the need to find a balance between cooperation and competition. Succeeding in an environment like this is facilitated by social abilities that require a certain kind of intelligence, such as the ability to manipulate others and get away with it (recall those primates who had perfected the art of deception).

At the same time, animals that are very cognitively complex tend to have a slow development (although there are exceptions to this, such as the common octopus, whose life expectancy in the wild barely surpasses one year, and yet they are exceedingly smart).[18] A slow development, in turn, favors sociality, for this allows the young to enjoy care and protection from their parents and other group members until they reach maturity.

Sociality is also often accompanied by high levels of EMOTION. As we have seen, complex social groups, especially those with fission-fusion dynamics, incorporate emotional incentives to pay attention to one's own and others' status within the group, which may favor death being something worthy of attention. Similarly, these groups tend to feature strong emotional bonds among different individuals, which may also incentivize their paying attention to others when they show signs of non-functionality.

Social animals are also often *K*-strategists, something which, as we already saw, favors a very strong affective bond between parents and their young. This may be crucial, for young animals represent the majority of those who die in nature, so this may be an "easy access" to the development of a concept of death. What is learned through the death of one's offspring—which will be more worthy of attention the more of a *K*-strategist an animal is—can be extrapolated to other cases that involve adult individuals, thus facilitating the emergence of a more general concept of death. Perhaps Evalyne or Tahlequah, after their experience with the death of their baby, would be more capable of rapidly processing what had happened when an adult member of their group passed away.

Last, sociality usually comes with high levels of EXPERIENCE. This is because social animals tend to live in social groups. Given that mortality rates in the wild are very high, especially among the youngest individuals, those social animals who live long enough will normally have plenty of opportunities to learn about death. Naturally, those animals who are social, but who because of personal circumstances live in an environment that shelters them from death, won't have an easy time learning about it. This is the case, for instance, for many of our companion animals or those who live in captivity and have access to veterinary care. When there's more than one animal in these environments, sick individuals are usually removed from the group to be euthanized, and the remaining ones are not given the opportunity to interact with the body and reach an understanding of what happened.

Despite sociality being a common companion of the holy trinity, it can't be considered as a necessary causal factor in animals' development of a concept of death, for not all social animals have high degrees of COGNITION, EMOTION, and EXPERIENCE. For instance, many species of insects are extremely social, but their members tend to be too cognitively simple to develop a concept of death. In contrast, what we see in eusocial insects tend to be stereotypical reactions to this phenomenon. Indeed, although among mammals sociality is a good predictor of cognitive complexity, in at least some insects we find the opposite to be true, and the size of the group in which the species lives is negatively correlated with the cognitive complexity that they show at the individual level.[19]

We can not only find sociality in the absence of the holy trinity, but we can also find the holy trinity in the absence of sociality. More specifically, as we will see in the following chapter, big predators are excellent candidates for the possession of a concept of death, but they are often not social animals, leading rather solitary lives instead. Sociality, therefore, is not a perfect indicator of where we are likely to find a concept of death. Although Bercovitch is right to remark that all thanatological studies to date have focused on social species, this is not due to an inescapable link between sociality and the concept of death, but rather stems from the emotional anthropocentrism that we saw in the previous chapter—that is, from the search on behalf of thanatologists of reactions to death that remind us of humans.'

Let's therefore stick to the holy trinity: COGNITION, EMOTION, and EXPERIENCE. With this in mind, we can begin to see why we can expect the concept of death to be very extended in nature.[20] To be clear, certain animals will be excluded from the start due to them not reaching the necessary base levels of any of these three factors. Still, we can predict that the concept of death will be relatively prevalent in the animal kingdom, for various reasons.

Emotion in the Wild

The first reason we can expect the concept of death to be widespread has to do with the EMOTION component, and everything that we discussed in the previous chapter. The concept of death is a *cognitive* achievement and isn't linked to any concrete emotion. If we were looking for grief in animals, we could expect the group of species capable of experiencing this emotion to be relatively small (though surely greater than has traditionally been assumed). This is because, for grief to be present, certain fairly concrete conditions must be in place that, as we saw, have to do with affection or love. In the case of the concept of death, we already pointed out that processing that someone has died can give rise to all kinds of emotions. But, in addition, all kinds of emotions may direct an individual's attention to corpses. In the same way that one can react in many different ways to the information that Babosín died, there can be many different reasons that can lead one to interact with a carcass before one has a notion of what death is.

An animal may pay attention to a corpse because she had a strong affective bond with the deceased, but also because she's very confused upon their lack of response, or because she feels curious about their new "attitude," or because she's scared or excited due to the immediately preceding events (an infanticide, someone's fall from a tree, the attack of a predator, and so on). Perhaps the body's lack of response or the odors that emanate from it lead to feelings of hunger, irritation, frustration, or sexual arousal in her. Perhaps the corpse's inertia is seen by her as a handy opportunity to demonstrate her dominance, or she might perceive it as a new and exciting toy in the environment. These emotions may give rise to different behaviors toward the carcass: tactile exploration, observation, sniffing, necrophilia, aggression, cannibalism, play. These different behaviors may not be performed with the *intention* to learn about death, but by entailing a temporally extended interaction with the inert body they may give the animal some crucial information about the corpse's irreversible non-functionality.

The plurality of emotions that may be involved in a learning process about death makes it easy for such learning to be fairly common, for it may in principle occur in animals with different personalities, intentions, relationships toward the deceased, and so on. Something similar occurs with the EXPERIENCE component: there is no concrete experience that is necessary for an animal to learn about death, and in the wild there are plenty of opportunities.

EXPERIENCE in the Wild

Humans have a tough time processing just how common and trivial death in the natural world is. For us, death is something that we know occurs but that we usually don't see. Perhaps because of this, different thanatologists and philosophers have defended that the concept of death is an abstract concept.[21] This makes sense from our perspective because many of us live in societies that keep us as sheltered from death as possible, and we tend to go through our whole lives with barely any opportunities to see corpses, much less to interact with them. Of course, this doesn't apply if you have a certain kind of job (forensic doctor or homicide inspector, for instance) or if you've had the misfortune to live through a horrible event, such as a terrorist attack or a natural catastrophe. Or if you're a serial killer with a host of victims buried in your garden.

Barring these exceptions, most of us reach the end of our lives having rarely come face to face with death. This is why it makes sense for us to think of it as something abstract. It also makes sense, from an anthropocentric perspective—which may be specific to the urban Global North—to think of death as an absence, for when our loved ones die, they're taken away from our sight, and we can only make sense of their deaths through the notion that they're *no longer there*. This is also why the capacity to understand the notion of absence, and even the notion of *absolute zero*, have been postulated as a prerequisite to reach a concept of death.[22]

For animals in nature, as well as for humans in certain societies, death is not something abstract, nor does it essentially consist of an absence. It's something very concrete and very tangible—something that can be smelled, touched, and tasted. The dead are not absent individuals, but essentially broken and irreparable bodies. It's important not to lose sight of this. The majority of animals' experiences with death will be very different from our own.

Not only are their experiences different, they also have many more opportunities to interact with the dead than we do. After all, death in nature is *everywhere*.

To begin with, this is due to the extremely high infant mortality rates in the wild. Any random animal will have a tough time reaching maturity, even among K-strategists, who have much higher survival rates than r-strategists. For instance, half of the chimpanzees of Mahale die before they are weaned.[23] Among the Serengeti lions, cub mortality may be as high as 67 percent.[24] And for r-strategists, these numbers can be astronomical. The Atlantic cod, for instance, can spawn millions of eggs at one time, but on average only one or two of the larvae that hatch will reach adulthood.[25]

An animal that survives its infancy doesn't thereby escape from the threat of perishing at any given moment, given the high prevalence in nature of factors such as predation, illnesses, parasites, accidents, natural disasters, lack of resources, intraspecific violence, and a host of causes related to the most dangerous of all species: *Homo sapiens*. In addition, one mustn't forget that different species don't live in isolation from each other, but rather share a habitat with many more. An animal who lives long enough will encounter death in a myriad of ways. It's true that many or perhaps the majority of these deaths will occur in individuals of other species over which the animal in question is not going to lose any sleep, but even in these cases the high presence of the EXPERIENCE factor may compensate the low presence of the EMOTION factor.

The EXPERIENCE and EMOTION factors of the holy trinity therefore won't pose any problem when it comes to giving animals access to a concept of death. But what about the COGNITION factor? In this case, things get a little more complicated, for we do need some pretty specific cognitive abilities for a concept of death to develop. However, as we shall see, there's no reason to think that it won't be within the reach of many nonhuman species.

As you may recall, in chapter 4 we established what an animal needs to be capable of processing for it to be warranted to attribute to her a concept of death: the bare minimum that she needs to be able to understand is that dead individuals don't do the things that living individuals of their kind do and that this is a permanent state. Or, to put it differently, the minimal concept of death is made up of the subcomponents of *nonfunctionality* and *irreversibility*. Let's look at these in turn.

COGNITION in the Wild: Non-Functionality

As we already pointed out, not all forms of non-functionality are relevant when it comes to thinking about death, as for example we don't want a simple rock (which is non-functional) to count as dead. Because of this, it's fundamental for the animal to be equipped with a minimal notion of life that allows her to identify *what is missing* in a dead individual. For her to be able to develop a minimal concept of death we thus need her to pay attention to the functionality *of the living* and for her to be able to contrast it with the non-functionality of the dead. And there are weighty reasons to think that this capacity is going to be very extended in nature. These have to do, on the one hand, with how common the ability to distinguish animate from inanimate entities is and, on the other hand, with how easy it is for animals to learn patterns and generate expectations from them.

The ability to distinguish animate from inanimate entities is exemplified by the capacity that we see in many vertebrate species to distinguish biological from nonbiological movement. In a series of experiments, Randolph Blake demonstrated this with cats.[26] He presented these animals with pairs of videos in which they could see points of light on a black screen. In one of the videos, the movement of the points of light corresponded with biological movement. These were points that were located as though they were on the head, torso, and limbs of a walking cat, and so their movement imitated that of the cat. In the other video, there was the same number of points, but these moved in a more or less random way. Blake trained the cats to select the video that corresponded to biological movement in order to get a reward and found out that they were amazingly good at it. They even managed to succeed in those cases in which the incorrect video showed points that only differed from the ones in the correct video in that their movement was asynchronous, but they appeared at the same heights and displayed the same movements as the video corresponding to the walking cat. The cats were also able to discriminate biological movement when the light points corresponded to a walking human. However, they failed when the video was upside down, that is, they showed a bias toward the natural direction of gravity.

This ability to distinguish biological movement appears to be highly conserved in the animal kingdom, having been demonstrated in species of mammals, birds, fish, and spiders.[27] In fact, it might even be innate in some species. In one study, newborn chicks were placed on a runway with a video projected on either side, each depicting points of light with biological or nonbiological movement, similar to the ones in Blake's experiment (see figure 10). The chicks showed a clear preference for the biological movement, even when it corresponded to a cat's and not a chicken's. What's most interesting about this is that the chicks had been bred in the dark and had not had any visual experience until that moment, which is evidence that this preference in them is innate and not learned.[28] And, once again, it doesn't manifest when the image is inverted and therefore doesn't follow the law of gravity.[29]

Be it innate or learned, many animals have the capacity to classify different entities in their environment as animate or inanimate, something that they do more or less spontaneously and which goes beyond their ability to distinguish different types of movement. Animals generate distinct *expectations* about an entity depending on whether they take her to be animate or inanimate, and they mistrust those entities that violate these expectations. If you live with a dog, you may have noticed that they find the vacuum cleaner and the hairdryer to be absolutely terrifying. This is because these are objects that violate

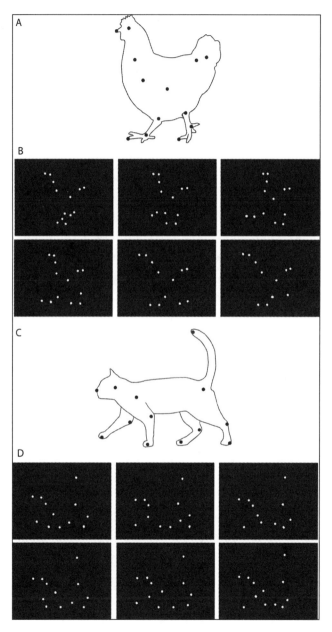

FIGURE 10. Animations depicting biological movement with points of light.

their expectations. By classifying them as inanimate objects, our pets don't expect them to generate the noise and movement that they do. That's why they find them so unsettling.

This effect has been demonstrated in experiments with various species. Jackdaws, for example, are suspicious of sticks that move on their own.[30] Infant Japanese macaques are wary of rocks that move in a self-propelled manner, an effect that is much less pronounced if the stone is covered with artificial fur.[31] Cats that hear a box that makes a noise when shaken expect an object to fall from it when it's turned over and, if it doesn't, they stare at the box for significantly longer (a sign that it has violated their expectations). This effect also happens in reverse: if the box doesn't make a noise when shaken, but an object falls out when it's flipped.[32]

Being able to distinguish animate from inanimate entities may be crucial for the survival of a species.[33] To begin with, because it allows an animal to tell its offspring apart from the rest of the elements in the environment and ensures that her care is directed toward them and not to any random rock or weed that may be lying around. But perhaps the most important reason why this distinction is so crucial has to do with the different degree of threat that is posed by animate as opposed to inanimate entities.

The first ones differ from the second ones in that they move in a self-propelled manner (without the necessary influence of external forces) and that their movement appears goal-oriented (that is, that they generally move *in order to* do something). This makes their movement more difficult to predict than inanimate entities' and, at the same time, much more important to monitor, especially because an animal's life (regardless of whether she's predator or prey) often depends on it. Consequently, animals tend to pay attention to any sign of animacy or goal-directed behavior in their surroundings, the importance of which is highlighted by the abundance of methods that species use to remain unnoticed, such as the mimicry of inanimate objects (think of those insects that look like leaves or sticks) or camouflage (which allows an animal to "blend in" with the inanimate background).

Animals' capacity to attribute different properties to animate and inanimate entities is nicely illustrated by an experiment carried out by Fuhimiro Kano and Josep Call with great apes.[34] They first showed them various videos in which they could see a human hand or a mechanical claw reaching out and grabbing one of two objects on a platform. Once the apes had habituated to these videos, the researchers used eye-tracking technology to follow the apes' gaze and see whether they anticipated the movement of the hand and the claw based on their previous experience. What they found was that apes anticipated the movement of the hand, predicting that it would grab the same object it had grabbed in the videos they'd already seen, even when those objects had switched places. In contrast, in the case of the claw their gaze was merely reactive, not anticipatory. That is, they observed its movement, but they didn't predict where it would go. Therefore, they only attributed goal-directed movement to the human hand.

In animals we also commonly see a capacity to discriminate patterns and develop expectations regarding how the entities in their surroundings typically behave, something that is crucial when it comes to having a notion of the functionality of living beings of this or that kind. We have seen that many animals feel uncomfortable when inanimate objects violate their expectations. This effect happens as well in the social realm. For instance, bonobos will protest more fiercely when they are the victims of aggression in a context in which they don't expect it than when the aggression corresponds to their social expectations.[35] Capuchin monkeys refuse to continue collaborating with experimenters when their conspecifics are getting a better reward than they are for performing the same task.[36] Chimpanzees who are given the option will choose to punish conspecifics who steal food from them.[37] Rhesus macaques who discover food but don't announce it to their group mates are likely to get a beating if caught.[38]

Expectations regarding how entities in one's environment typically behave are often what is called "cross-modal," which means that they implicate various senses. We saw this in the case of the cat that expects the box that makes a noise to contain an object. In this case, there is a *visual* expectation that stems from an *auditory* experience. Similarly, both dogs and cats will become wary if they see pictures of their human caretaker after hearing a recording of a stranger's voice, and likewise if the voice corresponds to their human but the picture depicts a stranger.[39] Dogs also act surprised if they encounter a different human from the one they expected after following an olfactory trail.[40] This "surprise effect" also occurs in horses who see a member of their herd go by followed by a playback of another individual's call.[41] And dolphins get weirded out if they hear a signature whistle that doesn't correspond to the conspecific whose urine they're currently tasting. Yes, you read that right.[42]

This ability to form cross-modal associations may be highly evolutionarily conserved. Tortoises associate high-pitch noises with small objects and low-pitch noises with larger ones, just as a human would do.[43] And bumblebees have been found capable of recognizing an object by touch when they have only seen but not touched it before, and vice versa.[44]

Having a notion of the *typical* behaviors and characteristics of the individuals in one's environment can be of great importance for survival, for it allows one to *predict* the behavior of friends and foes, as well as to detect *anomalous* behavior. A behavior is anomalous when it doesn't match what one expects from a specific individual or species. In order to recognize a behavior as anomalous, one needs to have a notion of the typical behavior of that individual or that species. In turn, anomalous behaviors are very important to monitor, for they are potential signs of a threat (such as a predator, a parasite, or a poisonous source of food) or an opportunity (such as prey that is easier to hunt down or an alpha male that can be deposed).

That animals have a notion of the typical behaviors of others is also exemplified by care behaviors exhibited toward individuals in distress. This is something that we saw, for instance, in the case of the elephant Grace, who was capable of perceiving that something was wrong with Eleanor and tried to help her stand upright.

We often see something similar in cetaceans. Kyum Park and collaborators, for instance, observed a group of long-beaked common dolphins trying to help one of their mates, who was dying.[45] It was a group of around twelve dolphins who swam very close to each other. One of them was moving strangely, with apparently paralyzed pectoral fins. The dolphins who were in the center of the group were helping her stay afloat by supporting her body and balancing it. Then they adopted a raft-like formation with their bodies, with the dying dolphin on top, which allowed her to continue breathing (see figure 11). When she at last died, the rest of her conspecifics remained next to the corpse, rubbing their bodies and heads against it to seemingly stimulate it.

The case of these dolphins, as well as Grace and Eleanor's, shows us that these animals have a notion of the typical behaviors of individuals of their species, for they don't usually go

FIGURE 11. A group of long-beaked common dolphins help a dying individual stay afloat.

through life helping others stay afloat or stand upright if they aren't displaying any kind of anomalous behavior.

Another relevant example is the special care that individuals with some form of disability sometimes receive. Sarah Turner, Lisa Gould, and David Duffus conducted a study in Awaji Island, in Japan, where a congenital malformation causes around 15 percent of Japanese macaques to be born with deformities in their digits or limbs that make it difficult or impossible to hold on to their progenitors.[46] These little monkeys' mothers adapt to their disability by holding them with one arm while they walk, despite this considerably affecting their movement capacities, as well as their balance and foraging, not to mention the fact that it entails a sizeable extra energy expenditure. In addition, these mothers breastfeed their young in a different way, for they actively ensure that they have access to the nipple and control the time the baby spends suckling.

Takuya Matsumoto and coauthors also reported on the case of a chimpanzee of Mahale who was born with a severe disability that affected her physical and neurological development.[47] Her mother was exhibiting the same kind of "compensatory care" that Turner and company had observed in the Awaji macaques. The mother was adapting to the limitations in the little one's abilities, as well as to her specific needs. She always carried her ventrally, holding her with one hand, offering her an extra point of support whenever they were sitting on a branch, and also making sure that she suckled from her breast. Moreover, the mother didn't allow nonrelated chimpanzees to carry the baby, even though she had allowed it with her previous offspring. Given that the other chimpanzees were showing no aversion toward her daughter, the authors speculated that this prohibition was probably due to the mother being aware that this infant needed special care and that it wasn't safe to let the other ones carry her.

The minimal concept of death doesn't require an explicit notion of life, but rather some implicit expectations regarding how individuals of a certain kind typically behave—expectations that will be violated by individuals who are sick or disabled, as well as by dead individuals. Given its importance for survival and care, it's to be expected that natural selection will have favored in many species the development of this capacity. At the same time, any animal who can develop expectations regarding how the animals in her environment typically behave will probably be capable of noting the total absence of behavior characteristic of corpses.

Having reached this point, it's important to highlight something that is not sufficiently emphasized in the thanatological literature, and that is how *radically different* a carcass is if we compare it to a live individual. A corpse isn't like a sleeping or unconscious individual. Even a fresh corpse *looks* completely different due to its absolute stillness, but will also *feel* completely different due to its total absence of responses to any interaction, as well as to its cold temperature or its limpness, or because of its *rigor mortis*; it also represents the total lack of *sounds* and will probably *smell* very different to those animals with a keen sense of smell. This is why it's so implausible to think that mothers who carry their dead infants for days or weeks won't have noticed that *something* is up with their baby, especially when we take into account that they can detect and adapt their caregiving to a partial non-functionality.

The non-functionality of death is therefore *multimodal*: it goes way beyond the mere absence of movement and incorporates signs that can be perceived with different senses. This multimodality appears to be exploited by many animals in their interactions with corpses.

For example, primates have been observed on many occasions insistently staring at the face of a conspecific's corpse.[48] Independently of whether this behavior is motivated by curiosity or by a natural tendency to fixate on faces, this part of the body in primates normally offers visual signs (facial expressions) with a communicative function, which presupposes that others of the same species can interpret them and makes it all the more likely that they can perceive their absence.

Tactile investigation also appears to be very important. I have already mentioned many cases of animals touching, stroking, holding, or grooming dead individuals, all of which are behaviors that will offer them perceptual cues about their state. Tactile interactions also include aggressive or sexual behaviors directed at carcasses. Fiona Stewart, Alexander Piel, and Robert O'Malley observed the case of an adult female chimpanzee whose corpse was treated with extreme aggression by the males of the group, who kept snatching it from one another and rough-handling it or dragging it across the ground.[49] The authors speculated that such aggressiveness could be due to their frustration upon the body's lack of response, or to them using it as a tool in their dominance displays toward the other members of the group. As for sexual behaviors, Giovanni Bearzi and colleagues point out that necrophilia, which is fairly common among cetaceans, can also be a way of displaying dominance, or that it can be a result of sexual arousal caused by the stress of the situation.[50] As in the case of the visual inspection of faces, both aggression toward corpses and necrophilia may not be carried out with the *intention* to investigate the non-functionality of a body, but they are behaviors that will give rise to very clear information regarding its lack of response.

The sense of smell is also used in the inspection of corpses. Cinzia Trapanese and collaborators tell us of the case of a capuchin monkey whom they saw closely inspecting the anal region of an infant who had been stillborn, and how immediately afterward he did the same with a live one who was in the surroundings,[51] presumably in order to compare the smells that emanated from both bottoms. For elephants, it's likely that smell is the most noteworthy characteristic of corpses, given their powerful olfactory abilities. In fact, they appear to be capable of distinguishing the bones of conspecifics from those of other animals using their sense of smell, and they have also been seen insistently sniffing the area where an elephant carcass has decomposed.[52]

A corpse's non-functionality can therefore be perceived through many different senses, and there are important reasons to consider that many animals will be capable of processing it. But, as we know perfectly well by now, the minimal concept of death is not just made up of *non-functionality*, but also of *irreversibility*.

COGNITION in the Wild: Irreversibility

On the face of it, it might seem as though requiring an understanding of irreversibility is going to drastically reduce the number of species that might reach a minimal concept of death, for irreversibility appears to be something considerably more difficult to process than non-functionality. In fact, as we already commented, human children have a tough time understanding that death is an irreversible process and tend to think of it as a merely temporal state.

However, we mustn't forget that the human experience with death is very particular and that it's not that easy to derive conclusions from it that can be extrapolated to other species. But, independent of our peculiar relationship with death, there seems to be something intuitive about the idea that processing irreversibility is more difficult than processing non-functionality. This is because irreversibility seems to incorporate a *temporal* element, for it is the notion that the dead individual won't exhibit those functions *ever again*. This has led some thanatologists to speculate that the concept of death requires the capacity to mentally time-travel or reason about the future.[53] Although, as we shall see, there's no reason to demand this type of cognition in order to process irreversibility, if we were to require it, we wouldn't necessarily be excluding nonhuman animals.

There are several studies that suggest that different animal species are capable of mentally traveling both to the past and to the future, remembering episodes they lived or planning for what may come.[54] Perhaps the most iconic studies in this regard are those that were carried out by Nicola Clayton and colleagues with Western scrub jays, a type of bird that lives in North America and is famous for her impressive capacity to remember the multiple places where she caches food. Clayton and company showed that they're not only capable of remembering where their hiding places are, but that they can also take into account when they used them to cache food and when they'll have access to them.

For instance, in one study the scrub jays were offered two types of food: one that they especially like but is highly perishable (worms) and another that they like significantly less but is edible for much longer (peanuts). The birds were given the option of storing both types of food for later recovery. After some time had gone by, they were given access to their hiding places. If just a few hours had passed, the scrub jays would go for the worms. In contrast, if several days had gone by, they would go directly for the peanuts, which showed that they were capable of remembering when they had done the caching and processing that it had been too long for the worms to still be yummy.[55]

Later studies showed that the jays are also capable of caching food with their future needs in mind. If they have learned that they might spend the night in a place where they'll be hungry the following morning, they will rather cache food there than in other places, as well as hide the specific types of food that they know will be missing there on the following day.[56] (Like when you go back to a hotel that you've already been to and you bring your own coffee because you've learned that the one they serve at breakfast tastes foul.) Scrub jays will also cache types of food of which they're satiated in that moment, anticipating that they might want to eat them once they gain access to their hiding places.[57] (Like when you go to the supermarket feeling full but you still grab some ice cream because you know, even though you don't feel like having any at the moment, it's highly unlikely to go to waste.) Therefore, they have the ability to predict their future needs, even when these are different from their current needs.

Although various species have demonstrated it, we still don't know for certain how extended the capacity to reason about the future is. However, this needn't worry us too much, for there's no reason to think that this type of complex reasoning is necessary to acquire a notion of irreversibility. In other words, we can understand the capacity to grasp irreversibility as an independent ability from temporal reasoning. This is because processing irreversibility is nothing but the *recategorization* of an animal from the class of individuals from which one expects the typical functions of her species to the class of individuals from which one does *not* expect these functions.

This allows us to distinguish how an animal would process the irreversibility of death from how she's likely to conceive of the non-functionality of an individual who is asleep. Sleeping animals don't exhibit many of the functions of living beings, but they're still animals of which one can *expect* these functions. That is, they're *reversibly* non-functional. An animal with a notion of what it means to be asleep won't be surprised if the guy who's having a nap suddenly gets up and goes for a stroll.

This absence of surprise is due to the existence of an *expectation of functionality*. An animal with a minimal concept of death, in contrast, *would* be surprised if the individual she had assumed to be dead suddenly moved, and, on the contrary, wouldn't be surprised if they didn't move again. Think of those poor lions who are left without an afternoon snack because the gazelle they had taken for dead has suddenly got up and skedaddled. If the lion is surprised, this is because she had an *expectation of non-functionality* that has been violated.

In a recent article, Arianna de Marco, Roberto Cozzolino, and Bernard Thierry argue that the evidence suggests that monkeys and apes are capable of processing that something is wrong with dead individuals, but they consider it more prudent to postulate that these primates are conceiving the dead as being in a state of dormancy.[58] According to this view, primates would not think of corpses as irreversibly non-functional, but rather as individuals of whom one can expect no behavior but who might get up and do stuff in the future.

However, there's no reason to think that this is a more parsimonious explanation. After all, it requires postulating two thoughts in the animal: that the other is likely not to do anything and also that she might do something. Furthermore, it requires the animal to have some grasp of probability notions. The reason de Marco and coauthors prefer this hypothesis to one that includes the *irreversibility* subcomponent is likely because they conceive of the latter in intellectualistic terms, as implying a notion of the absolute, of permanence, of "nevermore." However, as we've seen, there's no need to understand it in these terms—we can instead simply conceive it as the abandoning of an expectation of functionality in favor of an expectation of non-functionality.

Processing the irreversibility of death therefore doesn't require a capacity to reason about the future, but rather the ability to abandon one expectation in favor of another one. It doesn't require temporal or probabilistic reasoning, just the ability to revise the expectations that we have about others on the basis of our past experiences. There's nothing particularly sophisticated about this form of learning, so there's no reason to think that it can't be very present in the animal kingdom.

The multimodal character of corpses' non-functionality offers many opportunities for an animal's expectations to be violated and for her to start to process that the individual in question is not responding or behaving in the way in which individuals of her kind usually do. When it comes to grasping the irreversibility of this state, this will be supported by the fact that corpses have *their own functionality*, which is also multimodal.[59] Dead individuals don't just stop doing the things they did when they were alive; they also start doing new things—the things that dead individuals do. They start to smell in a strong and characteristic way, they bloat, they putrefy, they get covered in maggots, or they mummify and slowly dry up and disintegrate.

In some climates, these changes happen very fast. In the tropical forests of equatorial Africa, where many chimpanzees live, carcasses show signs of putrefaction a mere eight hours after death.[60] This may allow the individuals in the surroundings to have a clear perception that this is the same animal that was alive and well some hours ago. But even in those cases in which these changes occur more slowly, animals often spend their whole lives in the same habitat, and so corpses will follow similar change patterns. This may allow them to learn about the irreversibility of death, insofar as, once these cues appear, the individual never again shows signs of life.

Something relevant to study if we're looking for evidence of animals understanding the irreversibility of death is the contrast between behaviors directed at an individual while she's still alive and those she arouses once she's already dead. A good example of the relevance of this contrast comes from a heart-wrenching article recently published by Maël Leroux and collaborators.[61] In it they documented the reaction of a group of chimpanzees to an infant who had been born with albinism (an extremely rare condition in this species that had only been reported once before). When they encountered the mother with her baby, the chimpanzees appeared terrified, with their fur on end and emitting the same calls they use when they come across potentially dangerous animals, such as snakes or unknown humans. This fear contrasts with the curiosity and excitement that newborns tend to generate. After a few moments of panic, the alpha male finally snatched the albino infant away from the mother and, along with several other chimpanzees, attacked the baby until he passed away. Once he was dead, the observed behaviors radically changed. The majority of the group's chimpanzees were seen insistently inspecting the corpse, sniffing its head and its anogenital region, as well as stroking or pinching the fur on its back (see figure 12). They were apparently fascinated by this animal who smelled of chimpanzee but looked so strange. What's most interesting for the case at hand is that the group only gave in to their curiosity once the infant had died, which may be an indication that they knew the newborn posed no danger in that state, in other words, that his non-functionality was irreversible.

It's also important to monitor the context and the way in which animals abandon a body, something that has not received much attention in the thanatological literature. In fact, Sarah Brosnan and Jennifer Vonk complain that there's a bias toward

FIGURE 12. Two young chimpanzees tactilely and olfactorily inspecting the corpse of the infant born with albinism.

behaviors that are considered "interesting," while others that are "boring or commonplace" are not usually published, which leads to the perception that there is a higher proportion of interesting behaviors than there actually is.[62] The authors, for instance, refer to the case of Tahlequah, which could easily have been described in a scientific article and published, while no scientific journal would publish the case of a mother simply seen leaving her dead infant behind. And yet, while the authors rightly point out the importance of also documenting abandonment behaviors, they are wrong to consider them "boring or commonplace."

Let's consider for instance the case of the gorilla Simba, described by David Watts.[63] Simba had had several sons. Her firstborn, Mwelu, had died when he was eight months old, the victim of an infanticide. Simba had recently migrated into a new group, something that is always a cause for tensions, and little Mwelu died due to one of the resulting clashes. Her second-born survived, but her third and fourth offspring died due to illnesses. Watts recounts Simba's reaction to the death of her fourth infant:

> Simba had the corpse when I first saw her. She held it ventrally with one or both arms while resting and feeding and carried it ventrally, using one arm, when she moved. At 1320 h she sat to feed. Instead of cradling the corpse, however, she put it on the ground, in contact with her thigh, while using both hands to process plant stems. At 1325 h, she moved 2 m and again sat to feed. She left the corpse behind, the first time she had broken contact with it since I first saw it. She fed for 4 min, then returned to the body and briefly stood looking at it. She then took hold of its left foot with her right hand and pulled gently several times. Then she relinquished her grasp, looked at the infant for about 30 s, turned, and walked away. She did not return to the infant that day, nor did she or her group return to the area the next day. Other gorillas did not touch the infant or show particular interest in her body while I was following Simba, nor did any approach the body after she abandoned it.[64]

The way in which Simba abandons her infant is very interesting because of this apparent hesitation that she exhibits initially, followed by the last inspection of the body (as though to make sure that she's not making a mistake), and finally the resolution that she displays in her abandonment. She seems to have grasped that its non-functionality is irreversible. It's also interesting to consider Simba's particular story. She had previously had three offspring, and two of those had passed away. The factors EXPERIENCE and EMOTION were thus very present in her, which adds to the fact that as a gorilla she comes with high degrees of COGNITION.

Considering the context and personal history of each animal may provide important clues regarding their understanding of death. In general, we can expect there to be differences in the reactions to death depending on the past experiences of the individual. The younger an animal and the fewer deaths she's witnessed, the more we can expect her reaction to the sudden non-functionality of a conspecific to be more intense. In contrast, individuals who are already old and have experienced numerous deaths will tend to find it of little interest. This corresponds, for example, to what Arianna de Marco and colleagues found in the reactions of a group of tufted capuchin monkeys to the death of a groupmate.[65] The younger individuals, with few or no experiences with death, interacted most with the body, while the older ones, who'd had multiple experiences, barely showed any interest in it.

We must also take into account the reactions of those who *don't* have a close bond with the deceased. It's interesting, for instance, that apes and monkeys tend to pay little attention to the corpses of infants (with the obvious exception of mothers), despite how much interest these individuals arouse when they're alive. In fact, one of the arguments used to justify allowing great apes to reproduce in zoos is precisely that their life would be much more boring if there weren't any little ones around to give them joy and delight.[66] The fact that dead infants tend to be ignored by all except the mothers thus suggests that these animals can grasp their new state.

On occasion we see interesting reactions directed at the animal who is going through a grief process, which may also indicate an understanding of the situation. Zoë Goldsborough and coauthors, for example, describe the case of a group of chimpanzees that temporarily increased both the quantity and the quality of their affiliative behaviors directed at Moni, a chimpanzee who had recently lost her baby.[67] This increase took place even in individuals who had not had contact with her in the month before the little one's death, and who therefore weren't her close friends. The types of behaviors that they directed at her were those characteristic of reconciliation and consolation contexts, and which are often considered a sign of empathy. Most interesting was the fact that the individual who most increased her affiliation toward Moni was Tushi, a chimpanzee who didn't have an especially close relationship with the bereaved mother, but who had also lost a baby in the past. The authors speculated that her own experience going through a process of grief may have contributed to her empathizing with Moni, as is the case for humans.

There are therefore many reasons, both empirical and theoretical, to think that the minimal concept of death must be within reach for many animal species. Having reached this point, it's worth asking whether this is all that an animal may comprehend when it comes to death, or whether any nonhuman species could acquire a more complex concept. And the truth is that we can expect the concept of death to often emerge in nature in more than its minimal form.

The Natural Concept of Death

To understand how complex the concept of death might come to be in nature, we need to begin by remembering that the minimal concept of death is a *theoretical construct*. It's not a description of an empirical reality, but rather serves as a theoretical framework and as a benchmark to establish when we can say of an animal that she understands death. But the minimal concept of death doesn't have to be either a point of departure or a point of arrival in an animal's "mortal" learning process.

In fact, we can expect the concept of death to commonly emerge in nature with four subcomponents: *non-functionality, irreversibility, causality,* and *universality.* These last two subcomponents are not part of the minimal concept of death, but it's very likely that they will be, to some extent, present in the *natural* concept of death.

I say "to some extent" because an animal's understanding of the causality and universality of death can only ever be partial. In fact, it would be absurd to expect the opposite. A full comprehension of the universality of death would imply that the animal can not only process when something applies in a universal way—something that in itself is complicated—but also that she's capable of grouping all the living entities that she perceives (that is, con- and heterospecifics, as well as plants and fungi) under the same category as "things that can die." It seems unreasonable to expect any old animal to be able to do something like this, among other reasons because the majority of animals probably *couldn't care less* whether most of those beings are part of the class of things that can die.[68] Moreover, even in the case of us human beings, whom we like to think of as having a complete notion of the universality of death, we find difficulties establishing exactly which entities of the biological world can die and which can't. The never-ending discussion in biology with respect to the limits of the notion of "life" perfectly exemplifies this uncertainty.

Something similar occurs with the notion of causality. A precise and mechanistic comprehension of many of the common causes of death is something that, among humans, is restricted to a few experts, such as pathologists, and even among these there are high levels of uncertainty with respect to how some specific illnesses cause death. Therefore, it would once again be unreasonable to expect animals who don't have a way of accessing the necessary knowledge to have a complete understanding of the causality of death.

Still, it's more than likely that the animals who develop a concept of death will do so in a way that incorporates a *partial* understanding of its causality and universality. This is due to the prevalence in nature of two cognitive mechanisms: *association* and *inductive generalization*.

Association or associative learning allows an animal to mentally link two events that occur simultaneously or very closely in time in such a way that the next time she perceives one she expects the other one to occur. This would allow an animal to associate the irreversible loss of functionality that she has seen in one of her mates with the immediately preceding event (for instance, her falling from a tree or encountering a leopard).

However, the capacity to associate two events in a very rigid manner isn't by itself very useful. The need to interact with exactly the same stimulus without allowing for any kind of variation in order for the association to happen doesn't make much sense in rich and complex environments such as those where we can find organisms who depend a lot on their cognition in order to survive. This is where inductive generalization comes in, a capacity that allows these associations to be much more flexible and for the animal to carry out predictions that extend beyond the specific case of the original stimulus. It would, for instance, be the capacity to predict that what happened to Fulano when he fell from that tree or encountered that leopard would also happen to Mengano if he fell from a tree or encountered a leopard. Again, this is a capacity that is very extended in nature and would allow animals to have a (partial) notion of the universality of death.

In fact, we have some indications that some animals correctly process the contexts associated with death and tend to actively avoid them, which suggests that they operate with a concept of death with some presence of the subcomponents *universality* and *causality*. The most characteristic example of this are crows. These animals are known for gathering in large groups and even circling the corpses of conspecifics, something that has often been interpreted—from an emotionally anthropocentric standpoint—as a kind of funerary practice. In an experiment carried out in their natural environment, Kaeli Swift and John Marzluff wanted to determine whether this behavior allows crows to learn something about the circumstances of death that would serve them in the future.[69] To do this, they studied the responses to crows in the area of Seattle to certain stimuli presented two meters away from a source of food. These included a person with a distinctive mask holding a dead crow, a stuffed red-tailed hawk presented in conjunction with the masked person alone or in the vicinity of the dead crow, and the masked person holding a dead rock pigeon.

The researchers found that the crows behaved aggressively toward the hawk or the masked person when these had been presented in conjunction with a dead crow. They also often recruited support from other crows in the vicinity. Moreover, they tended to avoid feeding in the area where they had seen these stimuli, even when they were no longer present, which indicates that they took dead crows to be a sign of how dangerous an area is. They also formed a memory of the masked person that they had seen holding a conspecific's corpse and were able to distinguish them from other people who were wearing different masks. These effects persisted even six weeks after the stimuli had originally been presented.

Chimpanzees also show a sophisticated capacity to learn about the danger posed by different entities. For instance, the chimpanzees of Bossou, in Guinea, have learned that animal snares represent a grave threat and have on multiple occasions been seen trying to break or deactivate them, at least twice with success.[70] These primates also know that snakes are dangerous, and not only do they avoid them, but they also make sure that their groupmates are aware of their presence by emitting alarm calls, a behavior that is more likely if the others have not yet seen the snake or weren't present during the initial calls.[71] Furthermore, they have learned that crossing roads poses a risk, not just due to the presence of cars or humans, but also because it entails leaving the security of the jungle. As a result, they carefully observe the area before crossing and coordinate their movements in such a way that the bigger and stronger chimpanzees are at the front and back of the line, while the females and younger ones adopt the safer positions in the middle.[72] Last, they have been observed licking or cleaning with leaves their own wounds, as well as those of conspecifics, and experimental studies have also shown that they display physiological signs of distress in the presence of wounded individuals.[73]

Therefore, even though the minimal concept of death, as a theoretical construct, is just made up of the subcomponents *non-functionality* and *irreversibility*, the natural concept of death will likely emerge with, in addition, the subcomponents *causality* and *universality*. This is important because, on its own, the minimal concept of death only allows an animal to process what has happened to an individual who has indeed died. The natural concept of death, in contrast, also allows her to *predict* what could happen to others who presently are alive *if* they encountered the same lethal causes as their dead mates. This raises the question that you probably have been asking yourself for the last few pages: can these animals also comprehend that they themselves will eventually die?

Can Animals Understand Their Own Mortality?

The capacity to associate death with different events and generalize beyond the lived cases to future ones could in principle also allow an animal to develop a certain notion of *personal mortality*. Indeed, it's not outlandish to think that an animal who saw Fulano, Mengano, and Zutano die after falling from a tree or encountering a leopard might reach the conclusion that that very thing could also happen to her if she herself fell from a tree or encountered a leopard. This might require a somewhat developed level of self-awareness or slightly more complex cognitive capacities, such as conditional or counterfactual reasoning. Thus, the subcomponent *personal mortality* may be within reach for some animals, but the potential candidates will certainly be smaller in number when compared to the minimal or the natural concept of death.

The main problem here though is that it's difficult to come up with anything resembling conclusive proof of such an ability in a nonlinguistic species. Reports of animals fleeing from predators or chewing off their limbs to escape a snare will not do, for these behaviors can be explained by far simpler (and much more reliable) mechanisms than an awareness of their own mortality, such as a fear of certain stimuli, or extreme hunger or thirst. So where could we look?

In the thanatological literature it's sometimes mentioned that individuals who are close to dying commonly separate themselves from the group and disappear before passing away. This is often interpreted by laypeople as a sign that these animals *know* that they're going to die and this is why they leave, but it's actually more plausible to think that they just feel weak and can't keep up with the group, or that they go off on their own looking for peace and quiet for the simple reason that they don't feel physically well. This appears to be a more parsimonious explanation than their thinking that they're about to die, which on the contrary might motivate them to seek out social support, as it often does in our case.

An alternative indication of animals' capacity to conceive of their own mortality could be gleaned, one might think, from reports of suicidal behaviors. However, in this case the evidence that we have is scant, ambiguous, or anecdotal.

We know that many animals, when they are subjected to extreme stress, carry out self-destructive behaviors. In a series of horrifying experiments conducted in the 1960s and led by Harry Harlow, rhesus macaques were raised in complete social isolation.[74] Among many other issues, the monkeys developed a tendency to attack themselves, hitting their heads against the walls of their cages, biting themselves, and causing a loss of digits or lacerations on their arms and legs, and even gouging out their own eyes.[75] Precisely because of the social isolation that they had experienced, it's to be expected that these monkeys lacked a concept of death, so they probably didn't carry out these behaviors with the explicit intention of killing themselves. On the contrary, it's more likely that they were simply a manifestation of the psychopathology that they had developed as a result of the experimental conditions (which, one should add, had no other aim but to see how mentally ill these poor monkeys would become).

There are descriptions of self-mutilations in animals who are locked up in zoos, as well as in laboratory rodents who have been administered stimulating drugs.[76] Cases have also been reported in domestic animals. Some horses, for instance, have shown a tendency to bite themselves, as well as kick and lunge at objects.[77] Among cats and dogs it's also relatively common to see compulsive self-licking, nibbling, or scratching, which in extreme cases can result in serious injuries.[78] Again, these cases are explicable as anomalous behaviors due to adverse environmental conditions or pathologies, and don't seem to be evidence of suicidal tendencies.

Somewhat more suggestive are cases of dolphins who have supposedly ended their own lives. At least two cases have been reported, both occurring in dolphins who were captured from the wild and held captive in inadequate conditions. One is dolphin Kathy, the "actor" who portrayed Flipper for the longest period of time on the TV series of the same name.[79] The other is Peter, a dolphin who participated in a series of bizarre experiments from the 1960s led by John Lilly, which involved the psychedelic drug LSD, zoophilia, and attempts on behalf of the experimenters to communicate verbally and telepathically with the dolphins.[80] Both Kathy and Peter ended their own lives by voluntarily stopping breathing, according to the witnesses.

In contrast to our own case, for dolphins breathing is not a reflex, but each inhalation is instead the result of a voluntary decision, which is why they have at their disposal a fairly easy means to end their own lives. According to Ric O'Barry, Kathy's trainer who later became an activist for dolphin rights, "these are self-aware creatures with a brain larger than a human brain. If life becomes so unbearable, they just don't take the next breath. It's suicide."[81] However, these are very anecdotal cases—one of them observed by John Lilly, who by that time had lost all credibility due to his eccentric scientific endeavors. So while we can't disregard these examples, we also can't consider them very reliable. And even if the dolphins had indeed died after deciding to stop breathing, it would be impossible to know if this was motivated by a general apathy or depression caused by their captive conditions or by an explicit and conscious decision to kill themselves.

In any case, the fact that we lack reliable evidence of suicidal tendencies in animals does not necessarily imply that they lack a notion of *personal mortality*. We mustn't forget that suicide is a very maladaptive behavior—with some exceptions, such as those eusocial insects who sacrifice themselves for the good of the colony—and so we can expect natural selection to have acted against it. In fact, it would be more plausible for a notion of *personal mortality*, combined with the notion of *causality*, to lead animals to actively *avoid* those things that they have learned are lethal. Expecting a notion of *personal mortality* to lead to the appearance of suicide may just be a manifestation of our emotional anthropocentrism.

Animals' understanding of their own personal mortality is therefore particularly hard to study empirically and, with attempts to decode animal communication still in their infancy,[82] we have to content ourselves with the old comparative psychology adage: "Absence of evidence is not evidence of absence." That is, the fact that conclusive proof is hard to come by doesn't entail that the ability isn't there. What we can quite confidently say, however, is that if an animal incorporates the subcomponent *personal mortality* into her concept of death it would likely, again, be just a partial form of this notion. It would be equivalent to the thought that one *can* die, but not that one necessarily *will* die sooner or later, which is how we usually understand personal mortality. This is because our concept of death also incorporates the subcomponents *inevitability* and *unpredictability*. These two subcomponents are, I would wager, probably too abstract and complex for a nonlinguistic animal to grasp. However, as we'll see in the final chapter, this need not be interpreted as an insurmountable abyss between our concept of death and that of other animals.

We have thus reached the end of this chapter, in which we have seen that the concept of death emerges thanks to the interaction of the three causal factors that constitute the holy trinity: COGNITION, EMOTION, and EXPERIENCE. This is what makes elephants such good candidates for developing a concept of death, and is likely behind the fact that we see in them such interesting thanatological behaviors. However, we need not assume that this concept will only emerge in species that are as cognitively complex as these pachyderms. On the contrary, we've seen that animals in the wild have multiple opportunities to learn about death, that a wide spectrum of emotions may help them along the way, and that the cognition necessary to reach a minimal concept of death is both fairly unsophisticated and of great importance for survival, which inclines one to think that it must be very extended. In this chapter, I have also argued that the concept of death will tend to emerge in more than its minimal form, incorporating to some extent the subcomponents *causality, universality,* and in some cases perhaps *personal mortality.*

In the following chapter I will tackle one last aspect of death in nature that has been largely ignored in the thanatological literature: the topic of violence. As we will see, the ways in which violence takes place in the wild and the behaviors that surround it give us even more reasons to think that the concept of death will be very common among nonhuman animals.

7

The Opossum Who Was Both Dead and Alive

The Virginia opossum is a funny-looking animal. When captain John Smith first came across one, he famously said that she "hath a head like a Swine, & a taile like a Rat, and is of the Bignes of a Cat."[1] If Google had existed back then, the contemporaries of John Smith (he of Pocahontas fame) would have realized that he was exaggerating a tad, but we can fairly say that this is a species with a peculiar appearance: a kind of disheveled rat-racoon who permanently looks like she spent all night partying. Perhaps surprisingly, she has opposable thumbs on her hind feet and a hairless tail that she can use to hang from branches. She is also a marsupial. This means that, despite looking like a rodent, and not living in Australia, she carries her offspring in a pouch on her belly, like kangaroos.

But what is most remarkable about the opossum, and the reason I bring her up, is a behavior that she engages in when she feels threatened. If she senses that she has no chances of escaping, she becomes momentarily paralyzed, only to then fall on one side in something resembling the fetal position, with her tail curled up, her eyes and mouth wide open, and her tongue hanging out. In this pose, she stops responding to the world and starts to salivate, urinate, defecate, and expel a repugnant-smelling green goo from her anal glands. Her body temperature drops 0.6°C, her heart rate is reduced by 46 percent, and her breathing rate by 31 percent. Her tongue, usually pink, displays a blueish hue. In this putrefying-corpse state she waits, immobile, for the threat to pass. We could pinch her, or even cut her tail, and she would not react (this has actually been put to the test by somewhat psychopathic scientists).[2]

This little performance is so elaborate that the expression "to play possum" is used as a synonym for feigning death. And yet, despite how convincing her "costume" is and that she'd undoubtedly fool us if we didn't know her trick beforehand, the opossum in this state is still alive. From her absolute stillness, with her vital functions reduced to a minimum and emanating decay, the opossum is monitoring what is happening in her surroundings, ready to return to her daily routines as soon as the coast is clear. Not unlike the cat in physicist Erwin Schrödinger's famous paradox, we could say that the opossum is dead and alive at the same time.

The opossum's death display is the last piece of the puzzle that we're putting together in this book, and it is of crucial importance because it's one of the cases that best shows how common the concept of death is in nature. In this chapter, we will have a close look at the reasons why we can state this with almost absolute certainty. These relate to the fact that the opossum's behavior is a defense mechanism against predation that would not have the shape it does were it not for how extended the concept of death is among predators.

In order to understand this, we must begin by adopting a wider point of view and considering the role that violence plays in animals' development of a concept of death. For this, we shall examine the main manifestations of violence in nature and how they relate to a concept of death, a road that will lead us to predation and finally to the concrete example of the opossum.

Violence and the Concept of Death

As we saw in previous chapters, comparative thanatology has for the most part focused on both *intraspecific* and *affiliative* reactions of animals to death. That is, thanatologists have been searching for behaviors of care and affection that animals may direct toward dying or already dead conspecifics. The relation between violence in the animal kingdom and the concept that nonhuman animals might have of death has, in contrast, received very little attention. This is especially so in the case of *interspecific* violence, that is, the one that occurs among individuals of different species. The fixation on intraspecific and affiliative interactions, as we already saw, stems from the emotional anthropocentrism that permeates comparative thanatology, but this doesn't make it any less surprising, given how abundant violence is in nature and how, as we'll see, we find in violent contexts—and especially in predation—the perfect breeding ground for the concept of death to emerge.[3]

But let's take one step at a time. As discussed in previous chapters, on occasion animals interact with corpses in ways that are very far removed from anything that can be called caring and that may even be characterized as violent. Although these interactions are important because they offer the aggressor a lot of information about the irreversible non-functionality of death, what I'm interested in examining in this chapter is violence, not as a way of investigating the non-functionality of a body, but as the very cause of death.

In the wild, an important proportion of deaths take place as a result of violent interactions, some of which occur in intraspecific contexts (that is, within a single species). Animals often fight with conspecifics over their access to food, mates, or territory. These fights may end in the demise of one of the adversaries. In fact, in some species, a very high percentage of deaths are caused by another individual of the same species: among brown bears the number is around 10 percent; among long-tailed marmots, 15 percent; among meerkats (the species to which Timon from *The Lion King* belongs) it reaches a disturbing 19 percent.[4]

Often the deaths that occur because of intraspecific violence are accidental. We can, for instance, imagine a pair of stags caught up in a dominance fight that ends with the antlers of one of them severing an important artery of the other and causing him to die. In this case, we can assume that stag number one didn't intend to *kill* stag number two. And yet, we mustn't dismiss the importance that these types of accidents may have in the development of a concept of death in the animals involved.

These interactions are interesting, first, from the point of view of the aggressor, for having *caused* the other's death, even if it has happened accidentally, may make it a much more salient stimulus. Think of the difference in the attention that may be generated by a stag in the distance who succumbs due to old age and sickness and the one caused by the collapse of a stag who just moments ago represented a sufficiently strong and vigorous rival to be worthy of a dominance dispute. Being the causal agent behind the death may, in addition, contribute to its characteristics (especially the sudden non-functionality) becoming much more noticeable.

Second, this type of death may be especially useful for third-party witnesses to develop a concept of death. Violent interactions tend to attract the attention of group members—in the case of social animals—and deaths in these contexts tend to come with large wounds that present striking visual, olfactory, and tactile cues that allow other animals to learn about death.[5] The animal may, for example, learn to associate the color and brightness of blood, its metallic odor or its sticky, damp, and warm feel with the irreversible non-functionality that often follows it.

It may seem trivial, but we should emphasize at this point that one does not need a concept of death in order to kill someone. In fact, we can distinguish different levels of intentionality* or deliberateness, more or less close to a concept of death, depending on how far removed the killer's mental state is from the intention to kill the other (see table 2).

At the level zero we would find those cases in which there are no cognitive processes at all involved in the behavior that leads to another's death. For instance, let's imagine that Laura is watering the plants on her balcony. Suddenly, she's overcome by a monumental sneeze that causes her to drop the watering can. It falls from Laura's balcony and lands on the head of Lucas, who was smoking a cigarette on the street below, and kills him. Although it was Laura's sneeze that caused Lucas's death, her action possessed no intentionality, for a sudden sneeze is a purely mechanical process that doesn't involve the mental states of the individual in question.

At level one we can place those behaviors that, in contrast to a sneeze, are ruled by cognitive processes but don't imply any

---

* A note for readers familiar with philosophy of mind: I'm using the term "intentional" in its most colloquial sense, as a synonym for "on purpose" and not in order to refer to the "aboutness" of mental states. This is why I talk of *levels* of intentionality, rather than orders of intentionality à la Dennett. For all the mental states that I will discuss I assume first-order intentionality.

emotion or intention directed at the other, let alone a desire to hurt them. Let's suppose that Laura has attended a hypnosis show during which she has been conditioned to deliver a jump kick whenever she hears a klaxon. Let's imagine that she's happily watering the plants on her balcony when suddenly she hears Lucas sound his klaxon from the street below. When she hears this sound, Laura experiences an unbearable urge to jump kick. When she does, she hits one of the flowerpots on her balcony, and it falls to the street and splits poor Lucas's head open. In this case, there was a cognitive mechanism guiding Laura's behavior, since it's the result of her perceptually processing some information about the environment, but her action barely qualifies as deliberate because it reduces to a stimulus-response mechanism induced by a hypnosis session.

At level two we can find those behaviors that are under higher cognitive control, governed by emotions or intentions directed at the individual who ends up dead, but without the concrete and explicit intention to kill her. Imagine that Laura is playing *mus** with Lucas as her partner. Suppose they lose the game after Lucas's decision to go all-in in pairs with a couple of sixes.† Laura loses her patience and out of sheer frustration grabs an ashtray and throws it at Lucas's head, with such bad luck that he loses his balance, falls off the chair, breaks his neck, and dies instantly. In this case there was clearly an emotion directed at Lucas and an action that in principle could cause his death, but that is in no way done with that concrete intention. Although her action is voluntary, Laura didn't intend to *kill*

---

* A Spanish card game played by two teams of two players, where each team's fate strongly depends on the individual actions of each player.

† An extremely absurd move.

TABLE 2. Levels of intentionality in the behaviors that cause another's death

Laura example	Level of intentionality	Action caused by a cognitive process	Intention / emotion directed at the other	Intention to kill the other	Concept of death
Sneeze	0	No	No	No	Not necessary
Hypnosis	1	Yes	No	No	Not necessary
Ashtray	2	Yes	Yes	No	Not necessary
Poison	3	Yes	Yes	Yes	Necessary

Lucas, just to discharge (in a somewhat dramatic fashion) her rage at him.

At level three we would find those killings that are the result of an explicit intention to kill the other. Let's imagine that Laura, tired of always losing at *mus* due to Lucas's disastrous decisions, and incapable of telling him to his face just what an appalling player he is, decides to tackle the problem at the root and poison his food to get him out of the way and find a more suitable teammate. While in the case of her throwing the ashtray we would speak of manslaughter, in this case we would describe it as a murder, for Laura has the concrete and explicit intention to kill Lucas (in addition to acting with malice aforethought).

The presence of a concept of death is only necessary for those killings that have the highest level of intentionality. Although in the examples that we've been considering Laura is a human being, and thus we can presuppose that she has a concept of death, the latter is not *involved* in the first three cases. However, in order to have the *explicit* intention to kill someone, a concept of death is necessary. In fact, a minimal concept of death would not do, but instead we would require one with a certain presence of the subcomponents *causality* and *universality*, which would be necessary in order to choose the means with which to perform the assassination (in this case, a poison) and in order to predict that it will indeed kill the other individual (for Laura knows that Lucas belongs to the class of beings that can die).

It's normally assumed that animals who kill others possess second-level intentionality at most, since it's taken for granted that they lack a concept of death. Although in examples such as the stag fight that we were considering before there might be mental states directed at the other at the base of the behavior (such as certain emotions or the intention to defeat them in a fight), the explicit intention to kill the other would not be present. However, if the arguments I have defended until now are correct, the concept of death should be very present in nature, even often incorporating the subcomponents *universality* and *causality*. This might open the door to the possibility that some of the deaths we observe in the wild are the result of a behavior with an intentionality of level three. In this chapter, we will consider three contexts where evidence suggests that on occasion this level of intentionality does operate: coalitional attacks, infanticides, and predation.

Coalitional Attacks and the Concept of Death

Let's begin by considering those killings that are a result of coalitional attacks. This is a type of behavior that occurs most notoriously in our cousins the chimpanzees. In this kind of fight, two or more apes, usually male, ally themselves with each other to attack another, often with the intention of rising in the social hierarchy. Although coalitional attacks don't always culminate in the death of the individual who is assaulted, they offer some very suggestive evidence regarding the control these primates can have over their violent behavior. Consider this case narrated by Frans de Waal and that took place in the Arnhem Zoo, in Holland:

> Luit was alpha for only ten weeks. The Yeroen–Nikkie alliance made a comeback with a bloody vengeance one night during which the two allies together severely injured Luit. Apart from biting off fingers and toes and causing deep gashes everywhere, the two aggressors removed Luit's testicles, which were found on the cage floor. Luit died on the operating table due to loss of blood from the fight, which took place in a night cage with only the three senior males present. Given the victim's massive injuries and the relatively few injuries sustained by the other two, we must assume a remarkable level of coordination between Nikkie and Yeroen.[6]

Nikkie and Yeroen displayed a sophisticated level of coordination and violence that suggests that they had the explicit intention to hurt Luit. However, did Nikkie and Yeroen intend to kill him? It might be that in this case this intention wasn't there, for we're talking about captive chimpanzees who had likely had few or no direct experiences with death. However, we can perfectly imagine the different levels of intentionality that I sketched as admitting a gradation, which would allow us in this case to talk about a high level-two intentionality, for, even if they lacked the concrete intention to cause Luit's death, their intentions must have been very close to this one, if we assume that they were indeed trying to cause him as much harm as possible.

However, as we saw in the previous chapter, chimpanzees in the wild will have many opportunities to learn about death, and so we can expect that we will find here some killings with a level-three intentionality. This is precisely what is suggested by a case observed by Stefano Kaburu, Sara Inoue, and Nicholas

Newton-Fisher.[7] These primatologists observed a coalition of four chimpanzees attacking the alpha male of their group (named Pimu) until he died. The assault lasted more than two hours, during which Pimu was practically still—the authors speculate that he was in a state of shock. After these two hours, the attackers momentarily stopped, and Pimu emitted submission calls, a way of acknowledging that he had lost. However, the aggressors attacked him again, until at last he died. The authors write: "The escalation of violence beyond the necessary for [Alofu] to defeat [Pimu] remains difficult to explain. The attackers, in particular [Kalunde], seemed intent on ensuring [Pimu's] death."[8] Something similar to this determination to kill was observed by Toshishada Nishida, who found a defeated and dying alpha male and had to actively intervene to stop the rest of the chimpanzees of the group from attacking him further and killing him.[9]

Coalitional attacks, both in chimpanzees and in other species, may also be directed toward individuals who don't belong to the group, which is often the result of territorial disputes or the migration into the group of an external member. Lions, for instance, will attack any lion external to their pack who dares to enter into their territory, which often ends in the death of the stranger.[10] Wolf packs also have broad territories that they defend by means of howling and olfactory marks. These signs serve to warn other wolves not to enter, and this is how they are interpreted: a wolf that comes across one of these signs will feel intimidated and back away, unless hunger moves her to enter alien land, in which case she will not leave marks because she will be aware that the area belongs to another pack. Despite these precautionary measures, occasionally wolves of different packs do find each other, and death in these circumstances is so certain that it's considered one of the main causes behind the natural mortality of this species.[11]

Territorial disputes that end in death are not common only among big predators. Do you remember Marcel, the adorable little capuchin monkey that lived with Ross in the TV series *Friends*? Outside of television sets (where they absolutely do not belong), these monkeys are very aggressive and territorial and will launch a full-on attack toward any monkey that migrates into their land. Julie Gros-Lous, Susan Perry, and Joseph Manson report various cases of extreme coalitional aggression directed at solitary monkeys who tried to join an established group. The violence deployed once again suggests more than a mere intention to scare the intruder away. For instance:

> At 1344 hours, observers heard alarm calls from the newcomer and ran to investigate. Alpha male Pablo was chasing him on the ground, and then three other males [...] joined the chase, as the newcomer alarm called frantically. Several other males joined the chase on the ground, and they captured the newcomer and began biting him viciously, removing chunks of flesh. He screamed, roared and emitted alarm calls as he was torn up. The victim escaped briefly and was recaptured. Five males were observed biting his flesh. At one point two attackers bit the victim while three others held him down. About 3 min after the attack started, the victim broke free of his attackers and fled.[12]

The act of holding a victim down while others attack her is very interesting, as it suggests that the assaulters weren't merely following an individual aggressive impulse (which would presumably lead them to haphazardly beat up the intruder), but rather that they were in some sense *cooperating* with the other monkeys in that attack, suggesting some degree of understanding over what they were doing.

Such cooperative behavior is also characteristic of the coalitional attacks of chimpanzees. These apes, in addition, stand out among other species because their ambushes on extragroup individuals are not just opportunistic, but rather, groups of males often go on expeditions with the aim of finding and annihilating chimpanzees of neighboring territories. These fights occur when the assaulting group has numerical superiority, something that is relatively frequent, for the fission-fusion structure of these primates' societies favors the attackers finding chimpanzees who are alone or in small groups.[13]

Richard Wrangham and Dale Peterson describe the ambush of a group of seven males on a chimpanzee named Godi who belonged to another group and was alone at the time. As soon as he saw them coming, Godi tried to flee, but one of the attackers, Humphrey, grabbed one of his legs and made him lose his balance. While Godi was lying face-down, with his head pressed against the ground, Humphrey was holding his limbs, immobilizing him in a kind of armlock. The remaining chimpanzees then began to bite and hit Godi, while they shrieked and jumped, overwhelmed with excitement. They even hurled a big rock at him and left him severely injured, bleeding from dozens of wounds, and on the verge of death.[14] The authors describe these kinds of attacks in the following way:

> Defense of territory is widespread among many species, but the Kasekela chimpanzees were doing more than defending. They didn't wait to be alerted to the presence of intruders. Sometimes they moved right through border zones and penetrated half a mile or more into neighboring land. They did no feeding on these ventures. And three times I saw them

attack lone neighbors. So they seemed to be looking for encounters in the neighboring range. These expeditions were different from mere defense, or even border patrols. These were raids.[15]

Peterson and Wrangham comment that the brutality and apparent intentionality of these attacks is such that their discovery caused a huge stir in the field of ethology, for before that "[s]cientists thought that only humans deliberately sought out and killed members of their own species."[16]

Infanticide and the Concept of Death

Another context where we might find level-three intentionality—that is, the concrete and explicit intention to *kill* another individual—is infanticide. As we saw in previous chapters, surviving infancy if you're a wild animal is a heroic deed, and infanticide is partly to blame for this, since it's one of the most common causes of death among young animals of many species. Among hyenas, for instance, around 20 percent of cub mortality is due to infanticide.[17] In some chimpanzee communities, this number may be as high as 60 percent.[18]

Infanticide occurs when an adult kills a baby of her own species, which may or may not be her offspring. It's not a unitary phenomenon, so we can't find a single rule that explains all cases. Moreover, it's a very difficult behavior to study, given its often unpredictable character and because of ethical limitations, so we know very little about how it works and any discussion of the topic is necessarily going to be speculative. However, we do have some clues that indicate that, in at least some cases, a level-three intentionality is at the root of this behavior.

In order to see this, we need to begin by introducing a key distinction in ethology, which is the difference between ultimate and proximate causes. The ultimate cause of a behavior would be the reason why natural selection has favored its evolution, that is, the evolutionary advantage that comes with it. Proximate causes would be those mechanisms that make the behavior happen at a specific point in time, that is, whatever goes on in the environment or in the body of the animal that triggers that behavior.

This is easily understood with an example that at least some of you may be familiar with: having sex. If we ask ourselves why animals practice sex, there are two kinds of answers we can give, depending on whether we focus on ultimate or proximate causes. The ultimate causes of sex are the evolutionary benefits that the animal obtains from that behavior. In this case, it's pretty clear: sex brings with it the possibility of reproduction, so it's favored by natural selection because it contributes in a direct way to the individual's genes passing on to the next generation. But it's to be expected that most animals won't have reproduction in mind when they practice sex. That is, the proximate cause of sex is probably not a desire to reproduce, but some other more primary—and probably more reliable—mechanism, such as the search for sexual pleasure. Therefore, we could say that, from the perspective of the ultimate causes, animals practice sex *because this allows them to reproduce*, but also, from the perspective of the proximate causes, that animals practice sex *because it makes them feel good*. Both explanations are equally valid, for they operate at different levels.

Having said that, what happens with infanticide is that we have many hypotheses about its ultimate causes, but few ideas about what might cause it at the proximate level. But as we will see, in the case of at least some forms of infanticide, it's difficult to offer a proximate explanation of the behavior without invoking a level-three intentionality, that is, an intention to kill the infant. Let's therefore have a look at the types of infanticide that exist in nature.[19]

In the first place, we can find an accidental or arbitrary type of infanticide, the result of an instance of generalized aggression, as might happen in a territorial dispute among two groups of animals. In this kind of case, there needn't be a concrete intention to kill the infant, but rather the infanticide would often simply be the result of the little one being in the wrong place at the wrong time. However, even here we can find some suggestive examples.

Erin Vogel and Alexander Fuentes-Jiménez, for instance, tell us of the case of a group of capuchin monkeys who, in their attack on another group of monkeys, ended up cornering a mother with her baby in a river.[20] While the attackers were threatening her, the mother, half-submerged in the water, was tightly holding on to the infant while emitting alarm calls. While this was happening, suddenly the beta male of the group to which the mother in peril belonged appeared. When he was some fifteen meters away, he stood up tall and let out a threat call that made the attackers immediately turn around and start to chase him, thus leaving the mother and her baby alone.

What we see in this case is an altercation between two groups of monkeys and an instance of generalized aggression that could easily have resulted in the death of the infant. Although the assaulters seemed to have the mother and her baby as their goal in the beginning, the ease with which they changed their target suggests that they saw them merely as easy prey in their attack of that group. However, there's a chance that they may have had the intention of killing them, had they managed to grab hold of them.

In cases like this, in which there's generalized violence, it's difficult to determine whether there's a concrete intention to annihilate the infant or simply a general desire to cause as much harm as possible. However, it's worth pointing out that these types of episodes are infrequent. In most infanticides what we see is an aggression that (1) is clearly directed at the infant, (2) is not preceded by an episode of generalized aggression, and (3) immediately ends the moment the infant is eliminated. Due to this, Carel van Schaik has argued that what underlies these kinds of cases is a series of mechanisms that are distinct from mere aggressiveness.[21]

However, not all cases in which we see an infanticide happening as a result of directed aggression need be instances of level-three intentionality. Some cases may be the result of more or less stereotypical reactions brought about by stress. For instance, kangaroo mothers in danger have been observed taking their joey out of their pouch and throwing her far away while they flee, an action that allows them to reduce their load while distracting the predator.[22] When resources are scarce, some animals are also willing to eat one of their young, thus increasing their own and the other offspring's probabilities of surviving. Often those that are consumed are the ones that appear the weakest, whether because they are smaller or because they present some kind of imperfection. However, these cases are not too common. In fact, in the majority of infanticides the victim is not consumed.[23]

Leaving aside those infanticides that are brought about by situations of extreme scarcity or stress, as well as anomalous cases, such as that of the albino chimpanzee, who apparently was attacked solely due to his unusual looks, what reasons would an animal have to want to kill an infant? We can distinguish two central hypotheses: the sexual selection hypothesis and the resource competition hypothesis.

The sexual selection hypothesis establishes that infanticide is a behavior that males carry out in order to copulate with a female who in that moment is breastfeeding or caring for the offspring of some other male, and not sexually receptive. By killing her infant, the attacker increases the chances of passing on his own genes, since the mother becomes available for fertilization again.[24] This hypothesis operates at the level of the ultimate causes: it explains infanticide by appealing to reasons why it may be adaptive to kill an infant. Natural selection would have favored these behaviors because they tend to eliminate the descendants of sexual competitors at the same time as they increase the chances of the infanticidal male reproducing himself. But what would be the proximate cause of this type of infanticide? That is, what is going on in the mind of the aggressor? Might there be in this case a level-three intentionality—the intention, not just to attack the infant, but to end her life?

Some cases of infanticide for sexual selection seem governed by stereotypical mechanisms, which means that such cognitively demanding explanations are ruled out. This is what happens, for instance, in rodents, where infanticide obeys some pretty rigid rules that suggest that there's little intervention of cognitive processes. The males of these species tend to be infanticidal by default—the smell given off by pups seems to generate in them an irresistible urge to end them. This tendency is so strong that females have developed a surprising counterattack mechanism: if they're exposed to the smell of a strange male while they're pregnant, they're capable of reabsorbing the fetus so as to avoid wasting resources and energy on offspring that will be condemned to an early demise (this, by the way, is called the Bruce effect, a somewhat ironic name, given that the phenomenon was discovered by a woman: Hilda M. Bruce[25]).

However, when it comes to their own pups, male rodents seem to behave as any respectable father would, offering care and affection. This is not due to them recognizing the young as their own. In mice, for instance, this care is determined by a biological mechanism that inhibits infanticidal behaviors twenty days after ejaculation, which coincides with the length of the pregnancy in this species. Infanticidal behaviors spontaneously reemerge after two months, when the offspring have grown up and are safe from Dad's murderous tendencies.[26] In other rodents, such as gerbils, infanticidal motivations disappear once and for all after the male has looked after his first litter. However, the inhibition of infanticidal tendencies may be reverted if male and female are temporarily separated during the first pregnancy.[27] That is, if the female disappears for a couple of days in the period between copulation and birth, the male will kill the pups as soon as they're born. They thus seem to need both the smell and the uninterrupted presence of the female to give up the infanticidal life.

In other species, the proximate causes of this kind of infanticide appear to be much less rigid mechanisms. This is true in the case of primates. We have already mentioned that, among chimpanzees, infants tend to be an object of attraction and interest for the whole group, which includes the males, who therefore can't have such rigid infanticidal tendencies as rodents. In fact, infanticides tend to elicit very strong reactions on behalf of group members, who will scream in protest and try to intervene to stop it from happening. It has even been postulated that there might be a social norm against infanticide among these apes.

Claudia Rudolf von Rohr, Judith Burkart, and Carel van Schaik tested this hypothesis with an experiment in which they presented a group of chimpanzees with a series of videos in which they could see an unrelated group of chimpanzees carrying out four types of action: cracking nuts, attacking infants (including cases of infanticide), hunting colobus monkeys (which are of a similar size as chimpanzee babies), and fighting among adults.[28] What they found was that the chimpanzees spent significantly more time looking at the videos of aggressions against infants than at any other video.

This greater interest may be due to their being a social norm against infanticide in chimpanzee societies. Normally chimpanzee infants are the objects of huge amounts of tolerance on the part of the adults. In the words of Frans de Waal: "They can do nothing wrong, such as using the back of a dominant male as a trampoline, stealing food out of the hands of others, or hitting an older juvenile as hard as they can."[29] Consequently, the authors of this study interpreted the apes' interest in the infanticide videos as a sign that those images violated their expectations of socially acceptable behaviors.

The existence of a social norm against infanticide would only make sense if this were a behavior that was under the control of cognitive mechanisms, that is, if it were something that the chimpanzees could choose whether or not to carry out. And this seems to be the case, both for chimpanzees and for other primates, for infanticide seems to obey certain decision rules here, rather than rigid and stereotypical mechanisms like in the case of rodents. Male primates base their decision on whether to attack an infant on a whole set of considerations, as van Schaik describes.[30]

The first of these rules is of course the absence of paternity. Given that they likely don't recognize their own offspring, they base their decisions on the history of mating that they have with the infant's mother, in order to reduce their chances of killing one of their own sons or daughters. It appears that in this case the mere fact of having copulated is not enough to avoid an infanticide, but rather both the frequency of mating and the moment at which it occurred seem to play a role. Given the importance of this rule, in some primate species females tend to be very promiscuous. This way, they ensure that their babies' paternity can't be easily attributed to a concrete male, and thus they will be safer from possible infanticides.[31]

The second rule would be the chance of copulating in the future. Infanticidal males tend to be those that have recently become dominant or those that have high probabilities of becoming dominant. And this is a sensible rule, since in a community where low-ranking males never get a chance to mate, it wouldn't make sense for them to go around butchering others' offspring and taking their chances with the mother's and her allies' rage.

The third rule is connected to the age of the infant. Infanticides, as we already mentioned, tend to coincide with the periods of highest maternal investment, often lactation, for during these periods the mother isn't interested in having a fling with anyone. Van Schaik points out that in some species the younger individuals have a distinctive coat, which may serve as a stimulus that triggers infanticide. However, other stimuli may also do the trick, such as their smaller size or the attention they receive from the mother.

The final rule, and perhaps the most suggestive of a level-three intentionality, is the evaluation of immediate risks. Mothers usually defend their babies with all their might and can even recruit help from the father or other females of the group. However, in the majority of infanticides these defenses fail, and the attacker suffers no harm. This, in the words of van Schaik, is due to it being "a very unequal contest, where the would-be infanticidal male often is seen to follow the mother–infant pair for periods of hours, days or even weeks [...] and can simply wait for a brief moment of female inattention to attack the infant."[32] This suggests that infanticide is under the control of cognitive mechanisms, for the males choose the moment of attack with great care instead of simply responding automatically to certain cues such as the smell or the appearance of the infant. Moreover, it's rare for the mother to suffer any harm during these attacks, which would also suggest that the infanticidal male's concrete target is the baby.

Also suggestive of level-three intentionality are those infanticides carried out as a coordinated attack among two or more individuals. Jared Towers and collaborators describe the case of a male orca whose mother, already menopausal, helped him capture and kill a calf.[33] (In this species, females live for many years after they no longer reproduce, and they contribute to their progeny's reproductive success by acting as repositories of ecological knowledge.[34]) Mother and son chased the calf and her family for several kilometers, and when at last the son managed to trap her, the mother placed herself between him and the little orca's family, preventing them from saving the calf while her son strove to keep her underwater until she drowned. Even though the calf's mother tried with all her might to intervene, even strongly attacking the aggressor, her defense lasted for just a few minutes and stopped very suddenly, something that the authors relate to the rapid death of her baby. This case suggests both that the aggressors had the intention to kill the calf and that there was a fast recognition of the meaning of what had happened on the part of the mother. The authors describe the aggressors' behavior as "goal oriented because their chase led to an attack that was maintained until a particular outcome had been achieved. Once it had, their behaviour immediately changed."[35] The sudden change in attitude once they had managed their objective again suggests a level-three intentionality, that is, that they didn't just look to attack the calf, but to kill her.

The second big hypothesis that would explain those infanticides governed by cognitive mechanisms would be the resource competition hypothesis. The selection pressure in this case would be the higher availability of resources that comes with the death of an infant. It's a type of infanticide that, in contrast to the infanticide for sexual selection, tends to be perpetrated by females.

Dieter Lukas and Elise Huchard carried out an analysis of the factors that influence this type of infanticide.[36] The latter would be more probable in those species that live in places with adverse climate conditions that make the presence of food scarce or erratic. By killing another female's offspring, mothers would ensure more food for their own. Some females are also territorial and will kill the young of others who try to settle into neighboring regions.

However, the majority of these infanticides occur in species in which females live or raise their young together. On occasion, these infanticides happen when the offspring of another female attempts to suckle from the aggressor's breast. By killing the infant, the female would be defending (in a slightly excessive way) her own offspring's access to her milk. In those species that cooperate in raising their young, killing another female's offspring would also be adaptive, for it increases the proportion of care available for one's own. Furthermore, by killing others' babies, the mothers defend the future social status of their own, for they will have eliminated potential rivals.

All of these are explanations of infanticide from the perspective of the ultimate causes. But what occurs at the proximate level? Here again we can only speculate, but we do have some evidence that suggests that sometimes there might be a level-three intentionality. For instance, scientists have seen coordinated assaults on pups by groups of female hyenas. While some distract the mother by attacking her and keeping her away from her offspring, another female takes the chance to sneak in and kill them.[37] In some of these infanticides we can find apparent control and a concrete intention directed at the infants. Sarah Blaffer Hrdy describes as follows the attacks of dominant female wild dogs on subordinates' pups: "the behavior of the killer was systematic, goal-directed, and to an anthropocentric viewer premeditated, since [...] the killer waited for the opportunities to enter the burrow containing the pups unobserved by the mother."[38] This premeditation and goal-directedness once again points to a level-three intentionality.

Although infanticides and coalitional attacks are interesting for the topic at hand, they are also isolated events of relatively low frequency, something that may limit how much the perpetrators and witnesses can learn about them. Because of this, if we want to fully evaluate the importance of violence for the development of a concept of death in animals, it's crucial for us to consider one last context in which it happens in a much more common and frequent way: predation.

Predation and the Concept of Death

This planet is home to an immense number of predatory species with a very broad spectrum of detection, hunting, and subjugation techniques that in turn exemplify different levels of cognitive complexity. Therefore, we can't find a unitary rule that applies to all cases here either. However, we can use the table on the levels of intentionality (have another look at table 2) to carry out a somewhat rudimentary but hopefully helpful classification.

Some predators operate with level-zero intentionality. This would apply to many carnivorous plants. Pitcher plants, for instance, have leaves in the shape of small receptacles to which insects are drawn. Once they go inside one of these traps, they're not capable of escaping. At the bottom of these leaves there are some digestive juices that will dissolve the poor bug's body once she falls exhausted into them, thus allowing the plant to nourish from her. There's no kind of cognition guiding predation here, for the plant doesn't need to process any information or behave in any kind of way to feed on the insect; it is simply existing, and the process that leads to the death of the prey is a purely mechanical one.

At level one would be those predators that do operate with cognitive processes, but these are very simple and can't be said to be directed to the prey as such. This category could include those forms of predation that amount to stereotypical reactions to certain sensory stimuli that tend to signal prey. These are often "sit-and-wait" predators. Think of a toad who spends most of her day sitting still, waiting for an insect to appear that she can trap with her sticky tongue. The toad is probably not operating with a concept of prey, but rather is compelled to stick out her tongue whenever she sees a stain of a certain color that moves in a certain way. Her behavior implies some minimal cognition (she needs to process some information about the environment), but it resembles a stimulus-response mechanism, like Laura's tendency to jump kick when she hears a klaxon.

It could be argued that this level-one intentionality can be found in some carnivorous plants too, as with the case of the

FIGURE 13. This Venus flytrap is up to no good.

Venus flytrap.[39] This plant has some pincers that look like terrifying jaws in which it traps the insects that it feeds on (see figure 13). On the surface of these pincers there are some small hairs. When an insect lands on the plant and, in a lapse of fifteen to twenty seconds, touches one of these hairs twice, the pincer closes and the insect is trapped. The poor fellow will then start to wiggle in an attempt to escape, and with that movement she will touch those hairs even more. When she reaches three touches, the plant will begin to secrete digestive juices, and upon reaching five, the plant will start absorbing nutrients from its prey. Although it's not common to describe plants as beings with cognitive processes, this is such a sophisticated mechanism that it has led some authors to talk about a rudimentary numerosity and memory in this species.[40] Because of this, we might be able to grant this plant a level-one intentionality.

It is to be expected that the toad will exhibit somewhat more complex cognition than the Venus (for instance, if she has to calculate velocities, distances, and the exact moment in which to attack). Therefore, we might talk about a low level-one intentionality in the plant and a high level-one in the amphibian. In any case, both the toad and the Venus flytrap display a stereotypical reaction to certain stimuli that generally signal prey. They are very rigid behaviors and very dependent on concrete sensory stimuli. A human with some time to kill and dubious intentions could easily trick the toad with a video that featured stains with a similar aspect and movement as insects,[*] in the same way in which she could deceive the Venus flytrap by stimulating her hairs with a stick (or with her finger, if she feels intrepid). The fact that the toad and the Venus would continue displaying those reactions shows us that they're not operating with a concept of prey, in the same way as the necrophoric ants that we saw in chapter 2 weren't operating with a concept of death.

At level two we find those predators whose behavior is guided by emotions or intentions clearly directed at the prey as such, but who don't necessarily operate with a concept of death. Here we would mostly situate those predators who, in contrast to the toad, actively look for their prey, stalk them, and hunt them down. Predation also has an innate component here—the presence of a hunting instinct—but it's much more dependent on learning, which allows the animal to improve her stalking

---

[*] Such humans exist. See: https://youtu.be/sfSJP8avHWI.

and subjugation techniques. In addition, these predators tend to show great flexibility in how they carry out their behavior, and we see emotions in them that are clearly directed at the hunt itself.

The presence of these emotions relates to the fact that, for many species whose survival depends on them being successful in capturing and killing their prey, the hunt is linked to play, to enjoyment. This is something that you might have witnessed if you've ever lived with a cat. Maxeen Biben describes it as follows:

> [Predatory] behaviour patterns [...] show many properties commonly referred to as playful. [...] By play, I am referring to behaviour that is active, where the cat's attention is focused on the prey, or where the prey is touched, manipulated or approached, but not injured. These behaviour patterns can be considered unnecessary for predation since the prey could easily be killed with minimal preliminaries [...].[41]

Although this play, these "preliminaries," are not strictly necessary, they may have a very important evolutionary function. Associating the hunt with play and with joy gives the animal a very clear incentive to carry out a behavior that, as we shall see, can be very dangerous, but is absolutely crucial for her survival. After all, often those behaviors that contribute the most to our chances of surviving and reproducing are those we most enjoy—a simple but ingenious strategy that evolution has "come up with" to incentivize them.

In many hunting predators we thus see an intentionality of level two, that is, behaviors that lead to another's death and that are guided by mental states directed at the other, but that don't necessarily incorporate a concept of death—an explicit intention to kill the prey. For animals whose survival depends on hunting, this practice can't be dependent on something as arbitrary as their ability to develop a concept of death. In fact, although they clearly have a concept of prey and their predation is subject to complex cognitive mechanisms and learning processes, it's also normal for them to feel an irresistible attraction toward certain stimuli, which would serve to support their predatory behavior. When a cat goes crazy chasing a laser point, this is because it exemplifies a lot of what she finds attractive in her prey. It's a stimulus that she can't help but chase, even if it isn't reinforced with a feast at the end.

Although we can expect many predators to operate at the second level of intentionality, in the hunt and kill we find the optimal conditions for the concept of death to emerge, and it's more than likely that here we will often find level-three intentionality: the intention to cause the irreversible nonfunctionality of the prey. To see this, it's necessary for us to return to the holy trinity of the concept of death. If you recall, this is made up of the three causal factors COGNITION, EMOTION, and EXPERIENCE. In what follows, we'll see how these three factors are present to a high degree in many predators.

Let's begin by considering the presence of the causal factor EMOTION. As we've already mentioned, for many predators the hunt is linked to play. At the same time, the possibility of feeding on fresh meat often generates intense emotions. Dale Peterson and Richard Wrangham describe as follows the reaction of chimpanzees in this context:

> [T]he chimpanzees' visceral reaction to a hunt and kill is intense excitement. The forest comes alive with the barks and hoots and cries of the apes, and aroused newcomers race in from several directions. The monkey may be eaten alive, shrieking as it is torn apart. Dominant males try to seize the

prey, leading to fights and charges and screams of rage. For one or two hours or more, the thrilled apes tear apart and devour the monkey. This is blood lust in its rawest form [...].[42]

In some predators, this "blood lust" seems to transcend purely predatory contexts. This is the case, for instance, with some cetaceans. Steven Ferguson, Jeff Higdon, and Kristin Westdal interviewed a series of Inuit hunters about the predatory behavior that they had observed in the orcas in the area. What their interviewees described suggests that the hunt in this species often obeys motivations that go beyond hunger:

> Interviewees reported killer whales ramming narwhal to "break their ribs," and killer whales will "play with" the narwhal, or pieces of them, and throw them around. One interviewee provided a second-hand description of killer whales "playing soccer" with the narwhal, and another observed killer whales killing narwhal by slapping them with their tail. [...] One Pond Inlet interviewee described two killer whales biting a narwhal and pulling it apart, leaving the head and tail behind and taking the "meat in the middle." Two interviewees reported that killer whales will also drown narwhal. [...] Some participants noted that killer whales will sometimes kill and eat only a little of their prey, leaving the rest, and sometimes even not eat anything at all. [...] Eight interviewees noted that killer whales sometimes "kill for fun", kill without eating, or "play" with wildlife.[43]

These anecdotes suggest that, for orcas, attacking other animals—perhaps even killing them—is sufficiently fun to carry it out in those moments in which they're not hungry. (One can't help but being reminded of the joyful fellows of

Manganeses de la Polvorosa in Spain and their custom of throwing a live goat from the church's bell tower.*)

Dolphins exhibit similar behavior (if you wish to preserve your image of them as lovable and perpetually smiling animals, I recommend that you skip this paragraph). Mark Cotter, Daniela Maldini, and Thomas Jefferson (no, not that Thomas Jefferson) studied bottlenose dolphins' harassment and killing of harbor porpoises (a type of cetacean).[44] They observed that the dolphins used four different techniques in their attacks: (1) *sandwiching*, which consists of squeezing the porpoise between two dolphins, which often sends them flying out of the water and may fracture their ribs; (2) *drowning*, which entails holding the porpoise's head under water and preventing her from breathing; (3) *tossing*, or throwing the porpoise out of the water using fast and violent movements of the head or fluke; and (4) *ramming*, or running head first into the porpoise at high speed, often several times in a row or in conjunction with other dolphins.[45] These attacks, which frequently end up killing the porpoise, constitute a fairly common behavior that doesn't however correspond to predatory motives, for the dolphins never feed on their victims. Moreover, these don't seem to be territorial or defensive disputes, for porpoises tend not to frequent the same areas of the ocean as dolphins, nor do they attack them. Instead, the authors speculate that this is likely a form of play on the part of the dolphins; a way of testing out their fighting techniques and also bonding with the conspecifics with whom they carry out these attacks.

These examples show that predators tend to very much enjoy the hunt, to the degree that on occasion they carry it out as an

---

* This tradition, which used to be carried out as part of a winter festival, was banned in the year 2000.

end in itself. The factor EMOTION in predation is thus often very high. But, in addition, there are important incentives for the predator to attach great emotional value and thus pay a lot of attention to the specific moment their prey dies.

The first of these incentives derives from the bare fact that with the death of the prey comes a full stomach and the possibility of surviving one more day. In previous chapters, we saw that the demise of an animal may entail a huge loss for those closest to her, who may grieve for weeks or even months. This is the case, for instance, for those primates who carry their infant's corpse for extended periods of time, a behavior that, as we saw, has its roots in the affective bond between mother and baby. In the case of predators, the death of the prey is not seen as a loss, but the very opposite: as a gain. In this situation, death is a reason for joy. It is therefore to be expected that it will generate an emotion with the opposite valence as that of the mother who loses her infant, but not necessarily a less powerful one.

The second incentive to pay attention to the moment of death involves the extremely low success rate that hunters tend to have. Prey animals have as much interest in surviving as predators do in hunting them down. This has given rise to an arms race between predators and prey throughout evolutionary history. As a result, hunters usually have very developed muscles, excellent sensory systems, and are among the fastest animals on the planet. Prey animals, in turn, have developed sophisticated defense systems, which include advanced camouflage, deterrence, and flight techniques.

Depending on the efficacy of these defense systems, predators will have different levels of success, but they're still usually very low. A lion, for instance, has a 47 percent chance of trapping a warthog (a "Pumbaa" from *The Lion King*), 30 percent in the case of gazelles, zebras, and wildebeest, and 14 percent in the case of antelopes.[46] For red-tailed hawks, chipmunks are an easy prey, yet they only manage to catch 30 percent of those they chase. Cottontail rabbits are harder because they tend to zigzag when they run, which diminishes the chances of catching them to 18 percent. Grey squirrels, in turn, are their most evasive prey, because they have very fast reaction times and their flight movements are tridimensional and thus harder to predict. In this case, the hawks succeed a mere 12 percent of the time.[47]

The last big emotional incentive to pay attention to the death of the prey relates to the fact that live prey often represents a danger. Among prey's defense systems we can find some physical attributes, such as horns, claws, or sharp hooves, which in combination with aggressive tendencies can be very dangerous. One of the reasons why hawks have such difficulties hunting down grey squirrels is these rodents' notorious temper. Once caught, the squirrels aren't easily subjugated, but instead forcefully wriggle and try to scratch or bite their attacker.[48] Rats are also a hard prey to kill. In contrast to mice, whose main defense is running away and hiding, rats are capable of responding very aggressively and causing significant wounds to whoever is trying to hunt them.[49] Likewise, it's not at all rare for a lion to die at the hands (horns? hooves?) of a buffalo during a hunt.[50] In addition, a predator generally always risks being suddenly attacked by a prey who looks dead but is suddenly "revived." Because of all this, we can expect hunters to pay a lot of attention to the moment of death, for it allows them to let their guard down and enjoy the feast that awaits them.

Let's consider now the COGNITION factor. Big predators are often cognitively complex and the hunt is a behavior that, while being partly instinctual, also incorporates a learning process that allows the animal to refine her stalking and killing techniques.

The variety of methods that orcas and dolphins use to kill narwhal and porpoises shows us that this behavior is under cognitive control in these species. Lions also adjust their techniques to the size and danger that each species represents. Small prey are subjugated by means of a bite to their nape that tends to break their neck and kills them almost instantly. Larger prey, like zebras or wildebeest, are killed by means of a suffocating bite to their throat or snout. In these cases, the lion maintains her grip for however many minutes are necessary until the prey stops moving. The biggest and most dangerous prey animals, like giraffes and buffalo, are killed cooperatively among various lions.[51] Among domestic cats we can also distinguish the practice of playing with the prey from the killing act itself. Biben, in her study on predatory behavior in cats, comments that "killing was only rarely an 'accident' of vigorous play. In the great majority of cases, the killing bite was distinct from the preceding play activities."[52] This, again, points to a cognitive control, meaning the cat *chooses* when to kill the prey.

These examples suggest that in predation we can find level-three intentionality. But this level of intentionality requires a concept of death. What reasons do we have to think that the predator conceives the death of the prey as her switch to an *irreversible non-functionality*?

We have already seen some of the reasons. The importance that hunting has for their survival, their low success rate, and the danger posed by live prey are all incentives for them to pay attention to the moment of death, and specifically to the irreversible non-functionality that follows. Functional prey is prey that can escape and that can defend herself, so the predator has an interest in monitoring when the victim has become non-functional. In addition, some prey go into a state of shock when they're attacked, and stop moving; others—like the opossum—even actively pretend that they're dead. If for any reason the predator becomes distracted or lowers her guard, the prey may escape, so there's also an incentive here to learn about irreversibility. In addition, actively contributing to the disintegration of the corpse may allow them to learn that, after a certain point, there's no turning back for the prey.

As you may recall, the concept of death requires a certain notion of the vital functions that characterize individuals of a certain kind, for these are the functions that must be understood to terminate with death. Well, it turns out that there are also strong reasons to think that predators have a very clear idea of the typical functions that characterize their prey. This is because their low success rates, coupled with the importance that the hunt has for their survival, force them to be on the lookout for any sign of anomalous behavior or partial nonfunctionality in prey that may make them an easier catch. In fact, this seems to be what happens, for those animals who are very young or very old, who are ill, or have some kind of disability are much more likely to die at the hands of a predator than those who have all their typical capacities intact.

This phenomenon has been documented in numerous studies. Back in the 1950s, Lois Crisler studied the hunting habits of Alaskan wolves and verified that they have a difficult time catching healthy caribou. Even the calves are hard to catch if they're with the rest of the herd, who will gather around them to protect them. The majority of hunting scenes that Crisler managed to witness involved sick or disabled caribou.[53] In another, more recent study, in which the behavior of hunting dogs was investigated, the authors found that, though the higher vulnerability of certain prey is not always visually obvious to the human spectator, those animals that are caught exhibit a broad spectrum of deviations with respect to the typical condition of their bodies, mainly due to illnesses. Despite the prey animals having a strong incentive to pretend that they're healthy and well even when they're not, the authors speculate that these animals may have higher levels of stress that the dogs can olfactorily detect, which allows them to select the most vulnerable ones.[54]

Caroline Krumm and coauthors also verified that sick mule deer have a higher chance of being hunted down by a mountain lion. Although the illness makes the deer less alert and less fit, the authors consider that the mountain lions also learn to recognize and actively pursue these individuals.[55] This matches the behavior we see in lions. Throughout the day they see many potential prey, but they only chase a small proportion of them. According to George Schaller, who studied the behavior of this species in the Serengeti for many years, lions "undoubtedly" evaluate the possibilities of catching different animals according to their apparent vulnerability: "Most are given a glance, some merit a closer look, a few elicit hunting movements, and only a very few are actually pursued."[56]

Predators' fixation on the functionality of prey is such that some prey are capable of using it to their own advantage. For instance, some species of bird, such as the Kentish plover, like to pretend that they have a broken wing and can't fly, in order to attract the attention of the predator and lure her away from the nest (see figure 14). These are species that nest on the ground and whose nests are thus accessible to many predators.[57] By feigning this partial non-functionality, they postulate themselves as easy prey, thus capturing the predator's attention and preventing her from discovering the *real* easy meal. Despite how risky this behavior is, the offspring of these species have higher chances of surviving the more exaggerated their parents' performance and the closer to the predator that they dare to carry it out.[58]

FIGURE 14. A male Kentish plover pretending that he has a broken wing.

This is a very interesting behavior, for, despite being genetically inherited, it has a clear cognitive component. The bird never lures the predator toward the nest, but always away from it, so the behavior requires her to monitor the location of predator and nest at all times. In fact, the birds often tend to look in the direction of the predator and increase the intensity of their show if the target doesn't fall for it from the beginning.[59] Independently of the proximate mechanisms that generate this behavior in the birds, however, the point is that the existence of this strategy shows us that there's a tendency in predators to monitor the functionality of their prey, without which neither the birds' behavior nor its success would be explicable. This tendency to monitor functionality is what allows predators to detect the (apparent) partial non-functionality that the bird exhibits and categorize her as "easy" prey.

The last factor of the holy trinity, EXPERIENCE, is also present to a high degree in predation. A predator who survives to maturity will accumulate hundreds, if not thousands, of experiences with death. Although the majority of these will occur in individuals of other species, and thus there will be no affective bond linking her to them, for the reasons already specified these deaths will generate a lot of emotion and attention specifically directed to non-functionality and irreversibility. Even in the case of deaths that generate little emotion, the huge quantity of experience could easily compensate this relative lack of interest. After all, predators will have *daily* opportunities to learn about how their prey typically behaves, what characteristics make them easier to catch, which stalking and subjugation techniques work best, how to make the prey stop moving, and what signs indicate that their change of state is irreversible.

In predators we therefore find an extremely high presence of the three factors that make up the holy trinity of the concept of death. Much of what has been said applies to prey too, especially to those who live in big herds under the constant threat of the predators who inhabit the area. It's to be expected that they will have as much interest in not being caught as the predator has in catching them, if not more. They therefore have an important incentive to pay attention to the predator's behavior, and in fact what we see is that they're perfectly aware of the danger that each species poses. Schaller describes it thus:

> Prey animals know the potential of each predator intimately and react to such small nuances of behavior that my rather casual observations revealed only the most obvious signals to which they responded. For example, a gazelle has the smallest flight distance in response to a jackal and progressively larger ones in response to hyena, lion and leopard,

cheetah, and finally wild dog, which may be avoided as soon as a pack moves into view a kilometer or more away [...]. Similarly, wildebeest and zebra may permit wild dog and cheetah to approach to within 20 m or less without fleeing whereas a lion is usually avoided at 40 m. If, however, the herd contains vulnerable young, it may retreat when these predators are still 100 m away. Wildebeest and zebra are quite casual about the proximity of hyenas, permitting approach to within 10 m or less. Probably the animals can detect from the behavior of the hyena whether it is hunting [...].[60]

Prey animals therefore are not equally timid and cowardly toward all predators, but adapt their behavior to the danger that each species represents and the degree of vulnerability that the herd has in that moment. This is a sign that they pay attention to how each predator behaves and to the consequences that follow from their attacks.

A prey of this type who reaches maturity will have had the opportunity to see dozens of her groupmates succumb to predators. The EXPERIENCE factor will thus be very high. Fred Bercovitch, as we saw, points out that the thanatological literature doesn't include descriptions of reactions to death in prey that live in big herds, something that he thinks is to be expected given that these species tend to lack complex social systems, which presumably for Bercovitch implies that they lack reasons to have an emotional reaction to the death of their conspecifics.[61] However, this comment is made from the standpoint of emotional anthropocentrism. It's possible that for many species, staying in the area where there are predators hunting in order to express their emotions is perceived as too dangerous. At the same time, they may be so accustomed to their conspecifics dying that they conceive it for the most part as an event unworthy of much attention.

Still, it's not rare to see in prey protests against the destiny that awaits most of them. According to Schaller, mothers usually vigorously defend their young from the attacks of predators, efforts that extend to other members of the herd in those species that show higher social cohesion. However, these defenders will quickly flee if they see that the attackers are too numerous or too big and that their own life is in danger.[62] There's thus an unavoidable self-protection instinct that may often be strong enough to override altruistic motivations.

Given their importance, many of these fight or flight mechanisms will be governed by very primary mechanisms, such as the mother-infant bond or an innate fear of certain stimuli. However, the flexibility that we see here once again points to the intervention of cognitive mechanisms. Moreover, the accumulation of experiences makes it improbable for these animals to be absolutely clueless about the consequences that follow from the attack of a predator. This opens the door for the possibility that they might have a certain notion of their own mortality; of what could happen to them if a predator managed to catch them. In those species that are at the same time predator and prey, in which many of these factors converge, this possibility becomes even more plausible.

With this investigation of coalitional attacks, infanticide, and predation, we have seen that violence, both intraspecific and interspecific, is a fundamental place to look if we want to track the presence of a concept of death in nature. This is not just due to how easy it is to be killed by another animal in the wild, but also to the characteristics that tend to accompany these events and make them the perfect breeding ground for those involved to develop a concept of death.

Thanatosis and the Concept of Death

Now we're in a position to return to the example with which we began this chapter. As you will recall, the opossum is that bizarre—and arguably extremely cool—animal who disguises as the corpse version of herself when she feels threatened. This is a defense mechanism known as *thanatosis* that appears to have emerged independently several times in evolutionary history and that consists precisely of playing dead. Thanatosis is the last piece that we need to complete this puzzle because it's a mechanism whose very existence strongly suggests that the concept of death is very extended in nature.[63] This—and I cannot stress it enough—is not because the opossum herself *understands* that she's playing dead or does it *on purpose*, but rather because in order to explain the evolution of thanatosis we need to postulate a concept of death *in the deceived predators*. Let's have a look at the argument.

The first thing we need to do is distinguish thanatosis from another similar phenomenon—which may be evolutionarily related—with which it's often confused: so-called tonic immobility. Many species, when they feel threatened and with no chance of escaping, go into a kind of paralysis; a state of stillness that reduces their chances of being attacked or consumed by a predator. This is what is known as tonic immobility, and is found in a broad range of species, from insects to humans.[64]

While tonic immobility is a rather superficial behavior, which in some species may even be accompanied by an increase in heart rate,[65] in thanatosis the animal doesn't just stay still, but rather actively imitates the characteristics of a corpse. If you recall, in the opossum her vital functions and body temperature are reduced to a minimum; her bodily and facial expressions imitate those of a carcass; the animal urinates, defecates, her tongue turns blue, and her anal glands simulate the smell of rot. Although it's possible for thanatosis to have evolved from tonic immobility, it's much more than a mere paralysis: the animal is pretending to be dead. This is why, although the two terms are often used interchangeably, it's preferable to reserve the term "thanatosis" (from the Greek word *thánatos*, meaning "death") for these more ostentatious performances.

The opossum is perhaps the animal with the most elaborate and iconic thanatosis display, but she's not the only one to exhibit a behavior worthy of this label. Turkey vultures, when they feel threatened, lie belly down with their wings extended and their head against the ground. They remain absolutely still and don't react even if they're hit with a stick or lifted up off the ground. Howard Vogel tells the story of a hunter who thought he had killed one of these birds, put her in his sack, walked the two miles that it took him to get home, dropped the body in the backyard—apparently limp and lifeless—and when he came back out with his family to show them the haul, found her prying around the garden. As soon as she saw them coming, she played dead again.[66]

In frogs, thanatosis consists of staying still, generally with the eyes open, the limbs flaccid and extended and without responding to any interactions. Some species also stick out their tongue, and their breath may give off an ammonia-like smell (see figure 15).

Some snakes also engage in thanatosis (see figure 16). For instance, grass snakes lie belly up with an open mouth, their tongue hanging out, their eyes rolled, and exhibiting total limpness when picked up.[67] Gordon Burghardt tells us that hognose snakes precede their thanatosis displays with a dramatic "death," during which they writhe erratically, frantically, and violently, defecating and even biting themselves in the process. Following

FIGURE 15. Frog of the species *Acanthixalus spinosus* in thanatosis.

FIGURE 16. Grass snakes in thanatosis. The bottom right picture shows a snake that has been lifted off the ground and hangs limply.

this, they curl up their tail, lie belly up, and stay still with a gaping mouth, a hanging tongue, blood secreting from their mouths, and with no apparent signs of breathing. One may poke them and even lift them up and they wouldn't budge. The only mistake in this impressive show, Burghardt comments, is that they would immediately turn over if we tried to place them on the ground belly down.[68]

Tonic immobility is distinguished from thanatosis, not just because it's simpler, but also because in the majority of cases it fights off the predator in a different way, for it will tend to enter into play at other phases in the predatory sequence. To see this, we need to consider the different functions that prey defense mechanisms may have. As we have seen, predation has given rise to an evolutionary arms race, which has in turn favored the appearance of sophisticated defense mechanisms in prey (and their corresponding adaptations in predators). These mechanisms may act in any of the different predatory phases: (1) the detection of the prey, (2) the recognition of the prey as adequate, (3) the subjugation, or (4) the consumption.[69]

Anti-detection mechanisms act by avoiding the perceptual discrimination of the prey on the part of the predator. Here we can find behavioral tendencies such as avoiding areas or moments of the day in which it's more probable to run into predators, hanging around in large numbers to reduce per capita risk, or hiding as soon as a predator is detected. In this group we could also include all the techniques for blending into the environment that we see in prey and that prevent predators from perceptually distinguishing them.

Anti-recognition mechanisms do not block the sensory detection of the prey, but rather act by preventing the prey from looking edible or appealing. Here we can find, for instance, the mimicry of inanimate objects. Some insects are shaped like twigs or leaves, which doesn't keep the predator from seeing them (as in the case of camouflage), but rather makes them look like something different from the delicious little bug that they really are. Among anti-recognition mechanisms we can also find those that are meant to make the prey look bigger or more dangerous than she actually is, as well as *aposematism*, which is the use of warning signals, often by means of bright colors, which communicate to the predator that the animal is toxic and not fit for consumption. Some aposematic species resemble each other, a phenomenon known as *Müllerian mimicry*, which facilitates predators' learning to avoid them. The yellow and black stripes that wasps and bees share would be an example of Müllerian mimicry, which allows wasps to benefit from predators' having learned to avoid everything that looks like a bee, and vice versa. This phenomenon has given rise to some species who lack defense mechanisms mimicking the appearance of those who do, which is known as *Batesian mimicry*. Some flies, for instance, have yellow and black stripes despite the fact that they are non-toxic. They take advantage of the association between these colors and toxicity that some predators have formed due to their experience with other species.

Anti-subjugation mechanisms are those intended to prevent the predator who has detected the prey and identified her as something she wants to eat from actually capturing or subjugating her. Some prey species manage this simply by fleeing from the predator. This may include leaving behind a piece of their body to distract the predator and gain some time, like lizards do with their tail. Here we also find fight mechanisms. As we saw, many animals defend themselves by biting or scratching when they're attacked. Many of them are armed with horns, pincers, spines, or chemical defense mechanisms, like the skunk's infamous "farts." Prey animals may also opt for scaring or startling the predator by, for example, suddenly changing shape, displaying bright colors, or making a loud noise. Among anti-subjugation mechanisms we also find the distraction or attraction of the predator to lure her away from one's offspring, as we saw in birds who pretend to have a broken wing.

Last, *anti-consumption mechanisms* are meant to prevent the predator from eating the prey that she has already caught. Some species are toxic or venomous, others taste foul, and some are difficult to swallow for predators who can't chew. Anti-consumption mechanisms are the last line of defense, a final attempt to save one's life when all else has failed.

Returning to the distinction we're considering, thanatosis and tonic immobility fight off predators in different ways because they tend to operate in different phases of the predatory sequence. Tonic immobility is fundamentally an anti-detection and anti-consumption mechanism, whereas thanatosis is essentially an anti-recognition and anti-subjugation mechanism. Let's see how this works.

Tonic immobility sometimes operates as an *anti-detection* mechanism. This is because, by remaining still, the prey can better blend into the surroundings. It is thus a mechanism that contributes to camouflage. In addition, some predators only respond to movement, so tonic immobility would work by eliminating a key stimulus in attracting the predator's attention.[70] This technique can also help those animals that live in a group, for it increases the chances that other members who stay active will be the ones to catch the eye of the predator.[71] Some species are capable of using this technique in contexts outside of predation. For instance, young fire ants go into tonic immobility when an external group of ants invades their colony. Given that this species needs to perceive a kinetic signal in order to attack, by staying still the young ones ensure that the

FIGURE 17. Frog of the species *Phylomedusa bahiana* in tonic immobility.

adults, who are physically better prepared, are the target of any beatings.[72]

In addition, tonic immobility may also act as an *anti-consumption* mechanism. For example, in some toxic species of frog, tonic immobility consists of folding the limbs up against the body, which considerably reduces the animal's size (see figure 17). If one were to pull on a limb, instead of responding by going flaccid, it would fold back up against the body. This way, frogs protect themselves from possible wounds when they are consumed and they make it more likely that the predator will spit them out whole once she swallows them and verifies that they're toxic.[73]

FIGURE 18. Frog of the species *Pelophylax nigromaculatus* trying but failing to swallow a grasshopper in tonic immobility.

In other species, tonic immobility works by increasing the size of the prey, instead of reducing it. Some grasshoppers, for instance, prevent frogs from swallowing them by adopting a very rigid posture with their limbs extended (see figure 18). Since frogs don't chew, but rather swallow their prey whole, grasshoppers thus ensure that they won't fit in the frog's mouth. This technique also increases their chances of harming the frog's tongue or palate and thus being spit out.[74]

While tonic immobility has very clear defense functions, in the case of thanatosis, ethologists don't quite agree about the concrete advantages it offers and, thus, the reasons it evolved. Why in the world would an animal who wants to save her life pretend that she's already dead? There are different hypotheses, but all postulate thanatosis as an anti-recognition or anti-subjugation mechanism.

Thanatosis as an *anti-recognition* mechanism would work by making the prey seem unappealing in the eyes of the predator. This may be due to her having learned in the past that prey that have been dead for a while taste disgusting or to her having an innate aversion to decomposing corpses (so-called necrophobia). Rosalind Humphreys and Graeme Ruxton point out that, if it were to work this way, this would imply that predators aren't very smart, for thanatosis is usually initiated upon physical contact with the predator, and so the latter will have been able to see that the individual who now seems long dead was alive and well just a few seconds ago.[75] However, if what is operating here is indeed an (innate or learned) aversion to corpses, this needn't tell us much about the animal's smarts. Rejecting a prey in thanatosis that looked alive a second ago would be the equivalent of throwing out a can of beans after opening it and seeing that the contents are full of mold. Even if the can seemed completely normal at the start, and even if we lack a logical explanation for why its contents are rotting, we would be so disgusted upon opening it that we would throw it out without hesitation.

Although we can't rule out that thanatosis may sometimes work by generating disgust, it's not very plausible as the only explanation for why it evolved. This is because there are many ways of exploiting animals' capacity to feel disgust through much simpler mechanisms, like tasting bad or using chemical defenses. If thanatosis only aims to generate disgust in the predator, it's difficult to explain why it's so complex. The opossum, for instance, sometimes expels the putrid-smelling liquid from her anal glands on its own as a disgust-targeting mechanism.[76] Why the need, then, to stay still, reduce her vital functions, display a blue tongue, etc.? The opossum in thanatosis doesn't seem to be trying to appear disgusting, but to appear *dead*. (Again, this doesn't mean that she *intends* to look dead. Bear with me.) However, as we saw, many animals are born with a tendency toward necrophobia, a stereotypical aversion to carcasses that is not mediated by a concept of death. Could thanatosis work then by activating the predators' necrophobia, and could this be the reason why it takes the form it does? Again, it's possible for it to work this way sometimes, but it won't provide an explanation of thanatosis in all its complexity either. This is because necrophobia, as I explained in chapter 2, is linked to certain very concrete sensory stimuli, such as cadaverine or putrescine. These stimuli don't even have to be paired with a corpse for necrophobia to kick in. In a recent study, James Anderson, Hanling Yeow, and Satoshi Hirata found that chimpanzees will find a dead bird and a garden glove equally aversive if they are paired with the smell of putrescine.[77] The majority of the components of thanatosis would be irrelevant if it were a mechanism that was only "designed" to generate this stereotypical reaction. Thus, if this behavior evolved to exploit predators' necrophobia, we would expect it to be a necrophobia that was mediated by the concept of death. That is, the opossum's trick would work not just because she looks disgusting, but because she looks disgusting *because she appears dead* (and the predator has learned that dead individuals taste bad or upset the stomach, for example). In this case the predator's concept of death would be involved in the success of the mechanism.

Another way in which thanatosis could work as an antirecognition mechanism would be if it simply reduced the motivation of predators who specialize in live prey.[78] Perhaps a predator who wasn't particularly hungry and simply enjoyed the hunt would decide to discard the thanatosic prey, thinking that she was already dead or that she killed her, and would seek other victims. This would allow the prey to escape unscathed. Again, a concept of death would also be involved in the defense's success.

Some authors have postulated that thanatosis works, not as an anti-recognition mechanism, but as an *anti-subjugation* mechanism that would prevent the predator who has already identified the prey as an appetizing snack from actually killing her. The idea here is that thanatosis would relax the attention of the predator because she would think that she's already killed the prey, thus increasing the latter's chances of escaping without a scratch. Some predators treat their prey with relative gentleness or momentarily release them before eating them. Others cache their prey for a while after having killed them. Feigning one's death may allow an animal to remain unharmed in such situations.[79]

Patrick Gregory and collaborators point out that thanatosis as an anti-subjugation mechanism would have more chances of working in those cases in which there's a time lapse between catch and consumption. Therefore, it may also work against predators who try to kill multiple prey at one time, in such a way that their attention moves from one prey to the next before eating.[80] Something similar is suggested by Humphreys and Ruxton. For them, thanatosis will be more effective in those cases in which there's some kind of cost associated to ensuring that the prey is really dead. This may happen, for instance, when the predator has the opportunity to hunt down more than one animal, but there's only a short period of time during which the prey are within her jaws' reach.[81] These explanations presuppose that the predator has a concept of death that allows her to (erroneously) process that her victim's non-functionality is irreversible.

These hypotheses, however, would also have some difficulties in explaining the complexity of the more elaborate forms of thanatosis, for in principle it would be enough for the animal to stay still and limp—that is, in something close to tonic immobility—for the predator to take her for dead. All the extra paraphernalia that animals such as the opossum or the hognose snake add would seem unnecessary. However, it's precisely these extra pieces of "decoration" that point to the behavior looking to exploit the concept *of death* of the predators, for they amount to signs of both non-functionality and irreversibility. The stillness, the reduction of physiological functions, and the lack of reactions generate an illusion of absence of the organism's vital functions, while the secretion of blood, the disgusting smell, the blue tongue, and other signs suggest that the loss of functionality can't be reverted, contrary to what we would expect of a sleeping or unconscious individual.

Moreover, the fact that an animal in tonic immobility can also be taken for dead can serve to reinforce the argument that thanatosis aims to exploit predators' concept of death, if we think of the evolutionary relation between thanatosis and tonic immobility. Tonic immobility may indeed occasionally function as an anti-subjugation mechanism, for animals who engage in it often resemble a fresh corpse. It's perfectly plausible then that thanatosis evolved from that behavior, taking advantage precisely of the fact that predators may mistake an animal in tonic immobility for a dead one. The opossum and company would have specialized throughout their evolutionary history in perfecting the deathly look and increasing the chances of this error being made by predators.

It would, however, be fallacious to assume that we need to find a single hypothesis that can explain all cases of thanatosis in all their complexity. It's possible that there's more than one reason why thanatosis evolved, in the same way that it's conceivable for thanatosis to operate as an anti-recognition mechanism on some occasions and as an anti-subjugation mechanism on others throughout the lifetime of a single individual. Further, it has been speculated that thanatosis could sometimes work simply because some predators would be weirded out by prey who exhibit ambiguous signs of life and death, and would therefore prefer to leave the strange animal alone.[82] Thanatosis would thus be equivalent to other defense mechanisms that work by scaring or startling the predator. But this hypothesis would still require the predator to interpret the behavior as ambiguous signs of *life and death*, and thus would still need her to have a concept of death in order to work.

Although all of these issues are undoubtedly fascinating, what I'm interested in at this point is not solving once and for all what are the concrete reasons why natural selection favored the evolution of thanatosis as a defense mechanism. Instead, what I mean to argue is that *if* thanatosis has evolved, it's because there's some advantage—whichever form it takes—that comes from appearing specifically *dead*. We know that this is so because thanatosis is made up of a series of characteristics that are completely heterogeneous and whose only commonality is the fact that they're characteristics of the dead. What is there in common to lying belly up completely still, not responding to interactions, having one's mouth and eyes wide open, secreting blood from one's mouth, and defecating? What is there in common to having a blue tongue, smelling rotten, being cold, and not exhibiting signs of breathing? The only possible explanation as to why these characteristics occur *together* as part of the same behavior is that the animal is pretending to be dead.

This does not imply that opossums and hognose snakes necessarily have a concept of death and are behaving this way with the *intention* of being taken for dead. Instead, it appears to be a genetically inherited behavior that doesn't require learning and that they engage in automatically when they feel threatened. (Still, it's worth pointing out that we can see quite a lot of inter- and intra-individual variation in how they carry it out, and the moment in which they put an end to their little show seems to be under cognitive control.[83]) But independently of whether they are conscious that they're playing dead, what we can confidently assert is that the characteristics that thanatosis exhibits in species like these show us that the *selection pressure* that gave shape to them is their predators' concept of death. The opossum might not have a concept of death, but we can be pretty sure that at least some of those animals who intended to feed on her throughout evolutionary history did.

Behind a large part of the behaviors or characteristics that animals exhibit there is at least one selection pressure, which is what has favored their evolution. Occasionally, these selection pressures come from how their predators conceive the world, for this will determine whether a prey animal has a greater or lesser chance of being hunted down, and thus of surviving until reproduction.

For instance, those insects that are shaped like a leaf don't need to have a concept of leaf, nor an intention to resemble such an object. However, if insects with this appearance have evolved this is due to their predators not being interested in eating leaves. Throughout evolutionary history, the ancestors of these insects who most resembled leaves would have fewer probabilities of being eaten by a predator, and therefore more chances of reproducing, which in turn would make it more likely that succeeding generations would acquire a starker resemblance to leaves. The very existence of insects that look like this tells us something about the minds of their predators: that they tend to mistake them for leaves and that they do not desire to eat such things.

Similarly, the Batesian mimicry that we see in some flies is an indication of the associations that their predators can form. These flies don't need the *intention* to look like a wasp for their appearance to be beneficial. It's enough for their predators to have the capacity to form an association between the ideas *yellow-and-black-stripes* and *toxic-disgusting-blech*. This associative capacity would be the selection pressure that favored the evolution of these flies' Batesian mimicry. Thus, while this defense mechanism doesn't tell us much about the flies' mind, it does tell us something about the minds of their predators.

Likewise, and independently of the concrete evolutionary history behind thanatosis, this behavior gives us a window into the minds of predators. Thanatosis shows us that, for whatever reason, predators are less likely to eat those animals that they themselves *conceive* as dead beforehand. Throughout evolutionary history, those opossums who were more capable of mimicking death, and thus of tricking predators into thinking they were actually dead, had higher chances of surviving until reproduction, thus incentivizing a higher sophistication and complexity in this behavior in subsequent generations. Any explanation of the defense function of thanatosis needs to postulate a concept of death in the deceived predators, for only this selection pressure can explain why this behavior in these animals has the shape it has.

The very existence of thanatosis is thus a demonstration that at least some predators have a concept of death and that such a concept has an ecological impact. Even if it were "only" part of the minds of predators, it would have given shape to the behavior of other species.

That thanatosis is indicative of a concept of death in predators is also supported by the fact that this defense mechanism would not work against specialized predators, for these would have developed an appropriate response. Instead, it is to be expected that it works best against generalist predators who don't encounter these prey too frequently and are not familiar with their trick.[84] In turn, for it to work against generalist predators, these must have a *concept* of death, that is, not just the capacity to react to certain concrete stimuli that are associated with death, but a concept that integrates the holistic characteristics of death and which can be applied to different species. Only with a concept can a predator mistake for dead an animal that she has never seen before.

The distribution of thanatosis in the animal kingdom also points to how extended the concept of death is likely to be in nature. As we have seen, we find elaborate forms of thanatosis in, at least, amphibians, reptiles, birds, and mammals. Its widely extended, yet patchy, presence in the tree of life suggests that thanatosis appeared through convergent evolution in the different species that engage in it. These are species that are not closely related with each other, so it's unlikely for them to have inherited thanatosis from a common ancestor; rather, it's more probable that the presence of similar selection pressures in their different habitats caused this behavior to emerge independently in these different species. This, once more, points to the concept of death being ubiquitous in the animal kingdom, and thus far from a uniquely human feat.

8

Conclusion: The Animal Who Brought Flowers to the Dead

> We are a creature of organic substance and electricity
> that can be eaten, injured and dissipated back
> into the enigmatic physics of the universe.
> The truth is that being human is being animal.
>
> MELANIE CHALLENGER, *How to Be Animal*

There is an animal who buries her dead, adorns their tombs with sober gravestones or elaborate mausoleums, and brings them flowers on special occasions.

There is an animal who burns her dead, puts their ashes into urns, and decorates her living room with them or carefully scatters them in the deceased's favorite place.

There is an animal who wears mourning clothes and commemorates death dates; an animal who considers that one can also respect or wrong those who are no longer alive.

There is an animal who writes poems, theater plays, songs, novels, storybooks, and essays on death; an animal with a thousand words, symbols, and metaphors with which to refer to the Grim Reaper, the Dark Angel, the Banshee.

That our way of relating to death is unique in many regards is something that no one can deny. Throughout history, we have striven to emphasize these aspects and tell ourselves that we're the only animal who is conscious of mortality. Although our uniqueness is unquestionable, this emphasis is also a manifestation of our anthropocentrism. Whenever we take a uniquely human characteristic as that which settles the issue of which species possess X, it's to be expected—indeed, inevitable—that we will obtain as a result that our species is the only one that possesses X. Thus, if we were to take these uniquely human practices as the mark of an awareness of death, we would of course have to conclude that only humans possess this capacity.

In this book, I have argued that, if we want to tackle the question of the distribution in nature of a concept of death in a way that is fair, balanced, and—let's be honest—much more interesting, we need to leave aside both intellectual and emotional anthropocentrism. By doing this, we have been able to see that there are weighty reasons to consider that we're far from being the only animal with an awareness of mortality.

Having said that, it's also clear that we're the only animal with complex death-related rituals and symbolic representations of death—or at least the only extant animal, since evidence suggests that Neanderthals also engaged in rituals associated with death. How much does this matter? It depends on how anthropocentric we want to be in our judgment. For us, these rituals are therapeutic, healing; traditions that bring us together and give meaning, and for this reason they are crucial. For an ant who only cares about extracting from her nest everything that smells of oleic acid, all these activities would seem like an absurd way to spend time. For a vulture, our burying or burning our dead may appear like a terrible waste. And perhaps an opossum would feel pity for our inability to transform ourselves *ipso facto* into a cadaveric version of ourselves. The fact that these rituals are important *for us* doesn't justify our considering the concept of death that underpins them as the only true manifestation of mortality awareness.

However, we are probably the only animal with a notion of the inevitability and unpredictability of death. Could this establish an insurmountable abyss between our concept of death and that of other species? To my mind, this would be an implausible conclusion, due to the fact that the majority of us go through life paying little attention to the fact that we will inevitably die, that *each day* could be the day of our demise. This idea appears so horrifying that we push it to the back of our minds and we prefer to pretend that death is something that happens to others, not to us.

We have such a hard time accepting our own mortality that we have turned the topic into a taboo. It's considered bad taste for a woman to openly talk about the miscarriage she suffered. It's bad taste to manifest one's grief outside of the most private spaces. It's bad taste to say that someone is dying, even when everyone knows they're dying. If an old woman who's well into her nineties tells us about how she's arranging everything for when she's gone, we say: "Come on, don't talk about such things." If someone confesses to us that they have a terminal illness, we are left without words. Even those sentences that we tell ourselves to find consolation upon someone's demise ("At least she's resting," etc.) seem in this context too macabre.

We also don't like to see ourselves as the predators we are. The carcasses that city dwellers in the Global North consume come packaged in hygienic containers without a drop of blood, cut into neat little pieces so that they don't remind us too much of the animal to which they once belonged. Slaughterhouses are in the outskirts of neighborhoods, conveniently hidden from our view. We eat ham, pork, beef, instead of the *remains* of an animal who once lived, breathed, felt.

Humans hide death as much as they can. On the contrary, for animals in the wild, this is a day-to-day reality that can't be avoided. Despite lacking an explicit notion of the inevitability and unpredictability of death, they live out these notions much more than we do. A lioness who would die if she didn't kill and a gazelle who lives under the constant threat of being devoured will probably experience the reality of death much more vividly in their day-to-day than most of us do.

We also mustn't forget that this book was written by a human. Although I have made the strongest of efforts to leave my anthropocentric biases aside, there are important limitations in my capacity to theorize about the concept of death of other animals, which come from the simple fact that I can't abandon my *sapiens* brain. It's more than likely that there are sensory and semantic dimensions in animals' relationship with death that we're not capable of even imagining and perhaps not of understanding either, and which will therefore have been left out of the pages of this book. We would again be engaging in anthropocentrism if we were to assume that only our concept of death can have semantic content that is lacking in other species'.

Despite the unquestionable differences between our species and the rest, we have also seen that there are many continuities. We're not the only animal who understands death, nor are we the only one who grieves, nor the only one who kills on purpose or for fun. Scientists have been trying for a long time to find a characteristic that will definitively separate us from the other species. So far, all candidates have fallen. Neither the use of tools, nor culture, morality, or rationality are exclusive of human beings. Nor is a concept of death. We're not a unique species. We're just another animal. And as such, we're bodies that work until a certain point, but end up irreparably broken. Perhaps if we come to terms with the fact that we're animals we may also reconcile with our own mortality.

Acknowledgments for the Spanish Edition

To my editor at Plaza y Valdés, Marcos de Miguel, who put his trust in me and gave me the freedom to write exactly the book I wanted to write.

To the Austrian Science Fund (FWF), who financed this research through the projects M2518-G32 and P31466-G32.

To Mark Rowlands, for a foreword that, as with everything he writes, hits the nail right on the head.

To Javier Miró, for his stylistic advice and his support in the last few years.

To Pepo, Laura García-Portela, and Laura Danón, for being my beta readers and greatly improving the final product.

To all the people who, in one way or another, supported me throughout the journey: Laura, Eike, Tomi, Flo, Mat, Mikex, Sara, Alice, Hannah, Susu, Julia, Asier, Cristian, Judith, Birte, my parents, and my brothers. Love you loads.

Acknowledgments for the English Edition

This book was originally published in Spanish in 2021. I have had the rare opportunity to translate it myself three years later, which has also allowed me to undertake some revisions in light of peer review, editorial feedback, and the inevitable development in my own thinking. I have attempted however for the text to remain as faithful as possible to the original—and of course it still contains all the Easter eggs I left for my friends and family to find.

This edition also needs some extra thanks.

To my agents at ACER, Roberto Domínguez and Fátima Amechqar, for their belief in my book and their efficiency in securing an English edition.

To two anonymous reviewers for Princeton University Press, for their unqualified support of the project and their constructive comments.

To my editor, Rob Tempio, for his enthusiasm and incredibly helpful suggestions for improving the text.

To Chloe Coy, for accomplishing in record time the seemingly impossible task of securing the rights for all the pictures.

To Karen Verde, for her conscientious and hawkeyed copyediting.

To everyone else at Princeton University Press who helped with production and marketing.

To Richard Moore, who was the first person to read the English version of the text and saved me from the embarrassment of many false friends. Thank you for never failing to have my back, encouraging me to sleep against the tide, and thinking that I'm always the right amount.

To all who have given me laughter, love, and support during the last few months, with special mention, in no particular order, to Lau, for many endless conversations; to Paloma, for life-saving help in times of crisis; to Cristian, for being the kind of office mate who has an inflatable palm tree next to his desk; to Kristin and Lori, for their mentorship and warmth; to Fer, for being a role model of *la vida Fernando*; to Sam, for his calmness in turbulent times; to Sophie, for encouraging my spontaneous side; to Matt, for the hugs; to Joaquín, *por lo del reír*; to Susan, for the free therapy; to Mat, for the games and the corvid wisdom; to Julia, for the pseudoscientific talks on Berlin summer nights; to Sara, for the kindness and inspiration; to Tomi, for celebrating my creative side; to Hannah, for unapologetically giving in to our guilty pleasures; to Susu, for sharing this road since god knows when; to my parents, who gave me all my chances; and to my brothers, who share my sense of humor like no one else I know. Love you loads.

Notes

Chapter 1: Introduction: The Silence of the Chimps

1. Quoted in C. Irvine, "Chimpanzees' grief caught on camera in cameroon," *The Telegraph*, October 27, 2009. Available at: https://www.telegraph.co.uk/news/earth/wildlife/6444909/Chimpanzees-grief-caught-on-camera-in-Cameroon.html (accessed May 12, 2021).
2. See, for instance: Z. Reznikova and B. Ryabko, "Numerical competence in animals, with an insight from ants," *Behaviour* 148, no. 4 (2011): 405–434; G. Melis and S. Monsó, "Are humans the only rational animals?," *Philosophical Quarterly* (forthcoming); S. Monsó and K. Andrews, "Animal Moral Psychologies," in M. Vargas and J. M. Doris (eds.), *The Oxford Handbook of Moral Psychology* (New York: Oxford University Press, 2022), pp. 388–420; R. Moore, "Gricean communication and cognitive development," *Philosophical Quarterly* 67, no. 267 (2017): 303–326; C. Schuppli and C. P. van Schaik, "Animal cultures: How we've only seen the tip of the iceberg," *Evolutionary Human Sciences* 1 (2019): e2. DOI:10.1017/ehs.2019.1.

Chapter 2. The Ant Who Attended Her Own Funeral

1. This experiment is by E. Nowbahari et al., "Ants, *Cataglyphis cursor*, use precisely directed rescue behavior to free entrapped relatives," *PLoS ONE* 4, no. 8 (2009): e6573.
2. E. O. Wilson, N. I. Durlach, and L. M. Roth, "Chemical releaser of necrophoric behavior in ants," *Psyche: A Journal of Entomology* 65, no. 4 (1958): 108–114.
3. Including by myself; see S. Monsó, "How to tell if animals can understand death," *Erkenntnis* 87, no. 1 (2022): 117–136.
4. Q. Sun and X. Zhou, "Corpse management in social insects," *International Journal of Biological Sciences* 9, no. 3 (2013): 313–321.
5. Q. Sun, K. F. Haynes, and X. Zhou, "Managing the risks and rewards of death in eusocial insects," *Philosophical Transactions of the Royal Society B: Biological Sciences* 373, no. 1754 (2018): 20170258.

6. I have developed this topic in relation with the concept of death in Monsó, "How to tell," based on the analyses by C. Allen, "Animal concepts revisited: The use of self-monitoring as an empirical approach," *Erkenntnis* 51, no. 1 (1999): 537–544; H.-J. Glock, "Animals, thoughts and concepts," *Synthese* 123, no. 1 (2000): 35–64; A. Newen and A. Bartels, "Animal minds and the possession of concepts," *Philosophical Psychology* 20, no. 3 (2007): 283–308.

7. D. Davidson, "Rational animals," *Dialectica* 36, no. 4 (1982): 317–327.

8. This example originally appeared in N. Malcolm, "Thoughtless brutes," *Proceedings and Addresses of the American Philosophical Association* 46 (September 1973): 5–20.

9. Davidson's argument has been criticized in a myriad of texts, such as J. R. Searle, "Animal minds," *Midwest Studies in Philosophy* 19, no. 1 (1994): 206–219; Glock, "Animals, thoughts and concepts"; K. Andrews, "Interpreting autism: A critique of Davidson on thought and language," *Philosophical Psychology* 15, no. 3 (2002): 317–332; Newen and Bartels, "Animal minds"; A. Diéguez, "Conceptual thinking in animals: Some reflections on language, concepts, and mind," in J. Martínez-Contreras, and A. Ponce de León (eds.), *Darwin's Evolving Legacy* (Mexico: Siglo XXI and Universidad Veracruzana, 2011), pp. 383–395; M. Rowlands and S. Monsó, "Animals as reflexive thinkers: The aponoian paradigm," in L. Kalof (ed.), *The Oxford Handbook of Animal Studies* (Oxford: Oxford University Press, 2017), pp. 319–341.

10. K. Andrews, "Animal cognition," in E. N. Zalta (ed.), *The Stanford Encyclopedia of Philosophy*. Winter 2012. Available at: http://plato.stanford.edu/archives/win2012/entries/cognition-animal/

11. See K. Andrews, *The Animal Mind*, 2nd ed. (Abingdon, Oxon; New York: Routledge, 2020).

12. S. D. Preston and F.B.M. de Waal, "Empathy: Its ultimate and proximate bases," *Behavioral and Brain Sciences* 25, no. 1 (2002): 1–20; M. Tomasello, *A Natural History of Human Morality* (Cambridge, MA: Harvard University Press, 2016).

13. J.P.J. Pinel, B. B. Gorzalka, and F. Ladak, "Cadaverine and putrescine initiate the burial of dead conspecifics by rats," *Physiology & Behavior* 27, no. 5 (1981): 819–824.

14. Pinel et al., "Cadaverine and putrescine." This experiment did not include a control condition where the rats sprayed with cadaverine and putrescine were not anaesthetized, so I can't guarantee that these components in a non-anaesthetized rat would not trigger a burial attempt on behalf of her conspecifics. However, rats are much smarter than their popular image suggests, which makes it more than likely that this mechanism is not as rigid as the ants' necrophoresis. For an introduction to the social intelligence of rats, see M. K. Schweinfurth, "The social life of Norway rats (Rattus norvegicus)," *eLife* 9 (2020): e54020.

15. M. Yao et al., "The ancient chemistry of avoiding risks of predation and disease," *Evolutionary Biology* 36, no. 3 (2009): 267–281; G. S. Prounis and W. M. Shields, "Necrophobic behavior in small mammals," *Behavioural Processes* 94 (2013): 41–44.
16. A. Wisman and I. Shrira, "The Smell of death: Evidence that putrescine elicits threat management mechanisms," *Frontiers in Psychology* 6 (2015). DOI: 10.3389/fpsyg.2015.01274.
17. E.J.C. van Leeuwen et al., "Chimpanzees' responses to the dead body of a 9-year-old group member," *American Journal of Primatology* 78, no. 9 (2016): 914–922.

Chapter 3. The Whale Who Carried Her Baby Across Half the World

1. A. Selk, "Update: Orca abandons body of her dead calf after a heartbreaking, weeks-long journey," *Washington Post*, August 12, 2018. Available at: https://www.washingtonpost.com/news/animalia/wp/2018/08/10/the-stunning-devastating-weeks-long-journey-of-an-orca-and-her-dead-calf/ (accessed December 9, 2020); L. Pulkkinen, "Grieving orca mother carries dead calf for days as killer whales fight for survival," *The Guardian*, July 27, 2018. Available at: http://www.theguardian.com/environment/2018/jul/27/orca-mother-carries-dead-baby-washington-canada (accessed December 9, 2020); S. Darran, "Tour of grief is over for killer whale no longer carrying dead calf," *CNN*, August 13, 2018. Available at: https://www.cnn.com/2018/08/12/us/orca-whale-not-carrying-dead-baby-trnd/index.html (accessed December 9, 2020).
2. N. Wallington, "Tahlequah the orca—famous for carrying her dead calf for 17 days—gives birth again," *The Guardian*, September 7, 2020. Available at: http://www.theguardian.com/environment/2020/sep/07/tahlequah-the-orca-famous-for-carrying-her-dead-calf-for-17-days-gives-birth-again (accessed December 9, 2020).
3. J. Howard, "The 'grieving' orca mother? projecting emotions on animals is a sad mistake," *The Guardian*, August 14, 2018. Available at: http://www.theguardian.com/commentisfree/2018/aug/14/grieving-orca-mother-emotions-animals-mistake (accessed December 9, 2020).
4. D. J. Povinelli and T. J. Eddy, "What young chimpanzees know about seeing," *Monographs of the Society for Research in Child Development* 61, no. 3 (1996): i–vi, 1–152.
5. B. Hare, "Can competitive paradigms increase the validity of experiments on primate social cognition?," *Animal Cognition* 4, nos. 3–4 (2001): 269–280.
6. B. Hare et al., "Chimpanzees know what conspecifics do and do not see," *Animal Behaviour* 59, no. 4 (2000): 771–785.
7. J. Goodall, *The Chimpanzees of Gombe: Patterns of Behavior* (Cambridge, MA: Belknap Press, 1986).

8. V. K. Bentley-Condit and E. O. Smith, "Animal tool use: Current definitions and an updated comprehensive catalog," *Behaviour* 147, no. 2 (2010): 185–221; C. Brown, "Tool use in fishes," *Fish and Fisheries* 13, no. 1 (2012): 105–115; J. K. Finn, T. Tregenza, and M. D. Norman, "Defensive tool use in a coconut-carrying octopus," *Current Biology* 19, no. 23 (2009): R1069–R1070; M. Krutzen et al., "Cultural transmission of tool use in bottlenose dolphins," *Proceedings of the National Academy of Sciences* 102 (2005): 8939–8943; I. Maák et al., "Tool selection during foraging in two species of funnel ants," *Animal Behaviour* 123 (2017): 207–216.

9. R. M. Seyfarth, D. L. Cheney, and P. Marler, "Monkey responses to three different alarm calls: Evidence of predator classification and semantic communication," *Science* 210, no. 4471 (1980): 801–803.

10. The video is available at the following link: http://www.youtube.com/watch?v=Hh84Oe8JxUQ (accessed December 16, 2020).

11. S. Messenger, "Why this video of a beluga whale 'playing' with children is actually very sad," *The Dodo*, August 22, 2014. Available at: https://www.thedodo.com/why-this-video-of-a-beluga-wha-685343078.html (accessed December 16, 2020).

12. A. Whiten and R. W. Byrne, "Tactical deception in primates," *Behavioral and Brain Sciences* 11, no. 02 (1988): 233–244.

13. L. A. Bates et al., "Do elephants show empathy?," *Journal of Consciousness Studies* 15, nos. 10–11 (2008): 204–225.

14. C. Allen and M. D. Hauser, "Concept attribution in nonhuman animals: Theoretical and methodological problems in ascribing complex mental processes," *Philosophy of Science* 58, no. 2 (1991): 221–240.

15. A. Gonçalves and D. Biro, "Comparative thanatology, an integrative approach: exploring sensory / cognitive aspects of death recognition in vertebrates and invertebrates," *Philosophical Transactions of the Royal Society B: Biological Sciences* 373, no. 1754 (2018): 20170263.

16. Y. Sugiyama et al., "Carrying of dead infants by Japanese macaque (*Macaca fuscata*) mothers," *Anthropological Science* 117, no. 2 (2009): 113–119.

17. K. Andrews and B. Huss, "Anthropomorphism, anthropectomy, and the null hypothesis," *Biology and Philosophy* 29, no. 5 (2014): 711–729.

18. G. G. Gallup, "Chimpanzees: Self-recognition," *Science* 167, no. 3914 (1970): 86–87.

19. D. Shillito, G. G. Gallup, and B. Beck, "Factors affecting mirror behaviour in western lowland gorillas, Gorilla gorilla," *Animal Behaviour* 57, no. 5 (1999): 999–1004.

20. A. Horowitz, "Smelling themselves: Dogs investigate their own odours longer when modified in an 'olfactory mirror' test," *Behavioural Processes* 143 (2017): 17–24.

21. M. Kohda et al., "If a fish can pass the mark test, what are the implications for consciousness and self-awareness testing in animals?," *PLOS Biology* 17, no. 2 (2019): e3000021.

22. The distinction between intellectual anthropocentrism and emotional anthropocentrism was originally developed in S. Monsó and A. J. Osuna-Mascaró, "Death is common, so is understanding it: The concept of death in other species," *Synthese* 199, no. 1 (2021): 2251–2275.

Chapter 4: The Ape Who Played House with Corpses

1. J. D. Negrey and K. E. Langergraber, "Corpse-directed play parenting by a sterile adult female chimpanzee," *Primates* 61, no. 1 (2020): 29–34.

2. S. M. Kahlenberg and R. W. Wrangham, "Sex differences in chimpanzees' use of sticks as play objects resemble those of children," *Current Biology* 20, no. 24 (2010): R1067–R1068.

3. S. F. Brosnan and J. Vonk, "Nonhuman primate responses to death," in T. K. Shackelford and V. Zeigler-Hill (eds.), *Evolutionary Perspectives on Death* (Cham: Springer International, 2019), pp. 77–107; A. Gonçalves and S. Carvalho, "Death among primates: A critical review of non-human primate interactions towards their dead and dying," *Biological Reviews* 94, no. 4 (2019): 1502–1529.

4. Page 58 of A. De Marco, R. Cozzolino, and B. Thierry, "Prolonged transport and cannibalism of mummified infant remains by a Tonkean macaque mother," *Primates* 59, no. 1 (2018): 55–59.

5. Gonçalves and Carvalho, "Death among primates."

6. R. Cigman, "Death, misfortune and species inequality," *Philosophy and Public Affairs*, 10, no. 1 (1981): 47–64.

7. B. E. Rollin, "Death, telos and euthanasia," in F.L.B. Meijboom and E. N. Stassen (eds.), *The End of Animal Life: A Start for Ethical Debate* (Wageningen: Wageningen Academic Publishers, 2015), pp. 49–60.

8. C. Belshaw, "Death, pain, and animal life," in T. Višak and R. Garner (eds.), *The Ethics of Killing Animals* (New York: Oxford University Press, 2015), pp. 32–50.

9. B. Bradley, "Is death bad for a cow?," in T. Višak and R. Garner (eds.), *The Ethics of Killing Animals* (New York: Oxford University Press, 2015), pp. 51–64.

10. E. Harman, "The moral significance of animal pain and animal death," in T. L. Beauchamp and R. G. Frey, *The Oxford Handbook of Animal Ethics* (New York: Oxford University Press, 2011), pp. 726–737.

11. T. Regan, *The Case for Animal Rights*. Revised edition with a new preface (Berkeley: University of California Press, 2004).

12. Cigman, "Death, misfortune," p. 59.

13. Regan, *Case for Animal Rights*, p. 111.

14. Rollin, "Death, telos," p. 52. Heidegger's cited text corresponds to M. Heidegger, *Being and Time*. Trans. J. Stambaugh (Albany: State University of New York Press, 1996).

15. V. Slaughter, "Young children's understanding of death," *Australian Psychologist* 40, no. 3 (2005): 179–186; F. Bonoti, A. Leondari, and A. Mastora, "Exploring children's understanding of death: Through drawings and the death concept questionnaire," *Death Studies* 37, 1 (2013): 47–60.

16. M. de Unamuno, *Tragic Sense of Life*. Trans. J. E. Crawford Flitch (New York: Dover Publications, 1954), p. 38.

17. I originally introduced the minimal concept of death in S. Monsó, "How to tell if animals can understand death," *Erkenntnis* 87, no. 1 (2022): 117–136.

18. See Slaughter, "Young children's understanding."

19. Brosnan and Vonk, "Nonhuman primate responses"; Gonçalves and Carvalho, "Death among primates."

20. K. Andrews, "Chimpanzee mind reading: Don't stop believing," *Philosophy Compass* 12, no. 1 (2017): e12394.

21. M. W. Speece and S. B. Brent, "Children's understanding of death: A review of three components of a death concept," *Child Development* 55, no. 5 (1984): 1671–1686.

22. As suggested, as a marker for concept possession, by C. Allen, "Animal concepts revisited: The use of self-monitoring as an empirical approach," *Erkenntnis* 51, no. 1 (1999): 537–544.

Chapter 5. The Dog Who Mistook His Human for a Snack

1. M. A. Rothschild and V. Schneider, "On the temporal onset of postmortem animal scavenging: 'Motivation' of the animal," *Forensic Science International* 89, no. 1 (1997): 57–64.

2. T. Colard et al., "Specific patterns of canine scavenging in indoor settings," *Journal of Forensic Sciences* 60, no. 2 (2015): 495–500.

3. Colard et al., "Specific patterns."

4. D. S. Tuber et al., "Behavioral and glucocorticoid responses of adult domestic dogs (Canis familiaris) to companionship and social separation," *Journal of Comparative Psychology* 110, no. 1 (1996): 103–108; H. Chijiiwa et al., "Dogs avoid people who behave negatively to their owner: Third-party affective evaluation," *Animal Behaviour* 106, Supplement C (2015): 123–127; J. V. Bourg, J. E. Patterson, and C.D.L. Wynne, "Pet dogs (Canis lupus familiaris) release their trapped and distressed owners: Individual variation and evidence of emotional contagion," *PLOS ONE* 15, no. 4 (2020): e0231742.

5. J. Hughes and D. W. Macdonald, "A review of the interactions between free-roaming domestic dogs and wildlife," *Biological Conservation* 157 (2013): 341–351.

6. Colard et al., "Specific patterns."

7. J. Benz-Schwarzburg, S. Monsó, and L. Huber, "How dogs perceive humans and how humans should treat their pet dogs: Linking cognition with ethics," *Frontiers in Psychology* (2020): 11. DOI: 10.3389/fpsyg.2020.584037.

8. E. Engelhaupt, "Would your dog eat you if you died? Get the facts," *National Geographic*, November 12, 2017. Available at: https://www.nationalgeographic.co.uk/animals/would-your-dog-eat-you-if-you-died-get-facts (accessed January 7, 2021).

9. J. R. Anderson, A. Gillies, and L. C. Lock, "Pan thanatology," *Current Biology* 20, 8 (2010): R349–R351.

10. Anderson et al., "Pan thanatology," pp. R350–R351.

11. For an excellent philosophical analysis on human grief, see M. Ratcliffe, "Grief and the unity of emotion," *Midwest Studies in Philosophy* 41, no. 1 (2017): 154–174.

12. We can find a collection of animal grief stories in B. J. King, *How Animals Grieve* (Chicago: University of Chicago Press, 2013).

13. J. Goodall, *Through a Window: My Thirty Years with the Chimpanzees of Gombe* (Boston: Houghton Mifflin, 1990), p. 197.

14. A. Porter et al., "Behavioral responses around conspecific corpses in adult Eastern gorillas (Gorilla beringei spp.)," *PeerJ* 7 (2019): e6655.

15. A video of Segasira's behavior can be seen in the following link: https://figshare.com/articles/media/Mountain_gorilla_juvenile_male_Segasira_with_his_mother_Tuck_s_corpse_mp4/6198584

16. B. Yang, J. R. Anderson, and B.-G. Li, "Tending a dying adult in a wild multi-level primate society," *Current Biology* 26, no. 10 (2016): R403–R404.

17. I. Douglas-Hamilton et al. "Behavioural reactions of elephants towards a dying and deceased matriarch," *Applied Animal Behaviour Science* 100, nos. 1–2 (2006): 87–102.

18. Z. Muller, "The curious incident of the giraffe in the night time," *SWARA* 10, no. 3 (2010): 40–44.

19. D. De Kort et al., "Collared Peccary (Pecari tajacu) behavioral reactions toward a dead member of the herd," *Ethology* 124, no. 2 (2018): 131–134.

20. King, *How Animals Grieve*, p. 28.

21. King, *How Animals Grieve*, ch. 1.

22. For a review of deceased-infant carrying in primates, see: C. Watson and T. Matsuzawa, "Behaviour of nonhuman primate mothers toward their dead infants: Uncovering mechanisms," *Philosophical Transactions of the Royal Society B: Biological Sciences* 373, no. 1754 (2018): 20170261; S. Das et al., "Deceased-infant carrying in nonhuman anthropoids: Insights from systematic analysis and case studies of Bonnet Macaques (Macaca Radiata) and Lion-Tailed Macaques (Macaca silenus)," *Journal*

of *Comparative Psychology* 133, no. 2 (2019): 156–170; E. Fernández-Fueyo et al., "Why do some primate mothers carry their infant's corpse? A cross-species comparative study," *Proceedings of the Royal Society B: Biological Sciences* 288, no. 1959 (2021): 20210590.

23. R. Appleby, B. Smith, and D. Jones, "Observations of a free-ranging adult female dingo (Canis dingo) and littermates' responses to the death of a pup," *Behavioural Processes* 96, Supplement C (2013): 42–46.

24. For a systematic review of deceased-infant carrying in aquatic mammals, see M.A.L.V. Reggente et al., "Social relationships and death-related behaviour in aquatic mammals: A systematic review," *Philosophical Transactions of the Royal Society B: Biological Sciences* 373, no. 1754 (2018): 20170260.

25. A. De Marco, R. Cozzolino, and B. Thierry, "Prolonged transport and cannibalism of mummified infant remains by a Tonkean Macaque mother," *Primates* 59, no. 1 (2018): 55–59.

26. Watson and Matsuzawa, "Behaviour of nonhuman primate mothers."

27. K. A. Cronin et al., "Behavioral response of a chimpanzee mother toward her dead infant," *American Journal of Primatology* 73, no. 5 (2011): 415–421.

28. Watson and Matsuzawa, "Behaviour of nonhuman primate mothers," p. 9.

29. M.A.L. Reggente et al., "Nurturant behavior toward dead conspecifics in free-ranging mammals: New records for odontocetes and a general review," *Journal of Mammalogy* 97, no. 5 (2016): 1428–1434.

30. Y. Sugiyama et al., "Carrying of dead infants by Japanese Macaque (*Macaca fuscata*) mothers," *Anthropological Science* 117, no. 2 (2009): 113–119; D. Biro et al., "Chimpanzee mothers at Bossou, Guinea carry the mummified remains of their dead infants," *Current Biology* 20, no. 8 (2010): R351–R352.

31. C. Kingdon et al., "The role of healthcare professionals in encouraging parents to see and hold their stillborn baby: A meta-synthesis of qualitative studies," *PLOS ONE* 10, no. 7 (2015): e0130059.

32. A. Todorović, "My daughter came out. They handed her to me. She was dead," *Aeon* (2016). Available at: https://aeon.co/essays/my-daughter-came-out-they-handed-her-to-me-she-was-dead (accessed January 26, 2021).

33. R.S.C. Takeshita et al., "Changes in social behavior and fecal glucocorticoids in a Japanese Macaque (Macaca fuscata) carrying her dead infant," *Primates* 61, no. 1 (2020): 35–40.

Chapter 6. The Elephant Who Collected Ivory

1. K. McComb, L. Baker, and C. Moss, "African elephants show high levels of interest in the skulls and ivory of their own species," *Biology Letters* 2, no. 1 (2006): 26–28.

2. McComb et al., "African elephants."

3. J. Poole, "An exploration of a commonality between ourselves and elephants," *Etica & Animali* 9, no. 98 (1998): 85–110.

4. McComb et al., "African elephants."

5. McComb et al., "African elephants"; Poole, "An exploration"; J. Poole and P. Granli, "Signals, gestures, and behavior of African elephants," in C. J. Moss, H. Croze, and P. C. Lee (eds.), *The Amboseli Elephants: A Long-Term Perspective on a Long-Lived Mammal* (Chicago: University of Chicago Press, 2011), pp. 109–124.

6. The holy trinity of the concept of death was originally introduced in S. Monsó and A. J. Osuna-Mascaró, "Death is common, so is understanding it: The concept of death in other species," *Synthese* 199, no. 1 (2021): 2251–2275.

7. For an introduction to the communicative abilities of animals, see E. Meijer, *Animal Languages: The Secret Conversations of the Living World* (Cambridge, MA: MIT Press, 2020); T. Mustill, *How to Speak Whale: A Voyage into the Future of Animal Communication* (Townhead: William Collins, 2022).

8. J. R. Huntsinger, "Does emotion directly tune the scope of attention?," *Current Directions in Psychological Science* 22, no. 4 (2013): 265–270; B. Zikopoulos and H. Barbas, "Pathways for emotions and attention converge on the thalamic reticular nucleus in primates," *Journal of Neuroscience: The Official Journal of the Society for Neuroscience* 32, no. 15 (2012): 5338–5350.

9. For a review on the cognitive abilities of African and Asian elephants, see R. W. Byrne and L. A. Bates, "Elephant cognition: What we know about what elephants know," in C. J. Moss, H. Croze, and P. C. Lee (eds.), *The Amboseli Elephants: A Long-Term Perspective on a Long-Lived Mammal* (Chicago: University of Chicago Press, 2011), pp. 174–182.

10. C. J. Moss and P. C. Lee, "Female reproductive strategies: Individual life histories," in C. J. Moss, H. Croze, and P. C. Lee (eds.), *The Amboseli Elephants: A Long-Term Perspective on a Long-Lived Mammal* (Chicago: University of Chicago Press, 2011), pp. 187–204.

11. Moss and Lee, "Female reproductive strategies."

12. Moss and Lee, "Female reproductive strategies."

13. R. Appleby, B. Smith, and D. Jones, "Observations of a free-ranging adult female dingo (Canis dingo) and littermates' responses to the death of a pup," *Behavioural Processes* 96, Supplement C (2013): 42–46; G. Bearzi et al., "Whale and dolphin behavioural responses to dead conspecifics," *Zoology* 128 (2018): 1–15; F. B. Bercovitch, "A comparative perspective on the evolution of mammalian reactions to dead conspecifics," *Primates* 61, no. 1 (2020): 21–28; A. Gonçalves and D. Biro, "Comparative thanatology, an integrative approach: Exploring sensory / cognitive aspects of death recognition in vertebrates and invertebrates," *Philosophical Transactions of the Royal Society B: Biological Sciences* 373, no. 1754 (2018): 20170263.

14. Bearzi et al., "Whale and dolphin behavioural responses."
15. Bercovitch, "A comparative perspective."
16. S. Z. Goldenberg and G. Wittemyer, "Elephant behavior toward the dead: A review and insights from field observations," *Primates* 61, 1 (2020): 119–128; A. K. Piel and F. A. Stewart, "Non-human animal responses towards the dead and death: A comparative approach to understanding the evolution of human mortuary practices," in C. Renfrew and M. J. Boyd (eds.), *Death Rituals, Social Order and the Archaeology of Immortality in the Ancient World* (Cambridge: Cambridge University Press, 2015), pp. 15–26.
17. R. W. Byrne and A. Whiten, *Machiavellian Intelligence: Social Expertise and the Evolution of Intellect in Monkeys, Apes, and Humans* (Oxford: Clarendon Press, 1988); A. Whiten, and R. W. Byrne (eds.), *Machiavellian Intelligence II: Extensions and Evaluations*, 2nd ed. (Cambridge: Cambridge University Press, 1997).
18. P. Godfrey-Smith, *Other Minds: The Octopus, the Sea, and the Deep Origins of Consciousness* (New York: Farrar, Straus and Giroux, 2016).
19. S. O'Donnell et al., "Distributed cognition and social brains: Reductions in mushroom body investment accompanied the origins of sociality in wasps (Hymenoptera: Vespidae)," *Proceedings of the Royal Society B: Biological Sciences* 282, no. 1810 (2015): 20150791.
20. The majority of the arguments that follow were originally presented in Monsó and Osuna-Mascaró, "Death is common." Some were also sketched in S. Monsó, "How to tell if animals can understand death," *Erkenntnis* 87, no. 1 (2022): 117–136.
21. S. F. Brosnan and J. Vonk, "Nonhuman primate responses to death," in T. K. Shackelford and V. Zeigler-Hill (eds.), *Evolutionary Perspectives on Death* (Cham: Springer International, Evolutionary Psychology, 2019), pp. 77–107; B. E. Rollin, "Death, telos and euthanasia," in F.L.B. Meijboom and E. N. Stassen (eds.), *The End of Animal Life: A Start for Ethical Debate* (Wageningen: Wageningen Academic Publishers, 2015), pp. 49–60.
22. Brosnan and Vonk, "Nonhuman primate responses."
23. T. Nishida et al., "Demography, female life history, and reproductive profiles among the chimpanzees of Mahale," *American Journal of Primatology* 59, no. 3 (2003): 99–121.
24. G. B. Schaller, *The Serengeti Lion: A Study of Predator-Prey Relations* (Chicago: University of Chicago Press, 1976).
25. K. E. Jørstad, et al., "Atlantic cod—Gadus morhua," in T. Svåsand (ed.), *Genetic Impact of Aquaculture Activities on Native Populations* (2007), 10–16, p. 11. Available at http://genimpact.imr.no/__data/page/7650/atlantic_cod.pdf. Cited in O. Horta, "Debunking the idyllic view of natural processes: Population dynamics and suffering in the wild," *Telos: Revista Iberoamericana de Estudios Utilitaristas* 17, no. 1 (2010): 73–90.

26. R. Blake, "Cats perceive biological motion," *Psychological Science* 4, no. 1 (1993): 54–57.

27. J. Brown et al., "Perception of biological motion in common marmosets (Callithrix jacchus): By females only," *Animal Cognition* 13, no. 3 (2010): 555–564; T. Nakayasu and E. Watanabe, "Biological motion stimuli are attractive to Medaka fish," *Animal Cognition* 1, no. 3 (2014): 559–575; M. De Agrò et al., "Perception of biological motion by jumping spiders," *PLOS Biology* 19, no. 7 (2021): e3001172.

28. G. Vallortigara, L. Regolin, and F. Marconato, "Visually inexperienced chicks exhibit spontaneous preference for biological motion patterns," *PLOS Biology* 3, no. 7 (2005): e208.

29. G. Vallortigara and L. Regolin, "Gravity bias in the interpretation of biological motion by inexperienced chicks," *Current Biology* 16, no. 8 (2006): R279–280.

30. A. L. Greggor et al., "Wild jackdaws are wary of objects that violate expectations of animacy," *Royal Society Open Science* 5, no. 10 (2018): 181070.

31. S. Tsutsumi et al., "'Infant monkeys' concept of animacy: The role of eyes and fluffiness," *Primates* 53, no. 2 (2012): 113–119.

32. S. Takagi et al., "There's no ball without noise: Cats' prediction of an object from noise," *Animal Cognition* 19, no. 5 (2016): 1043–1047.

33. A. Gonçalves and S. Carvalho, "Death among primates: A critical review of non-human primate interactions towards their dead and dying," *Biological Reviews* 94, no. 4 (2019): 1502–1529.

34. F. Kano and J. Call, "Great apes generate goal-based action predictions: An eye-tracking study," *Psychological Science* 25, no. 9 (2014): 1691–1698.

35. Z. Clay et al., "Bonobos (*Pan paniscus*) vocally protest against violations of social expectations," *Journal of Comparative Psychology* 130, no. 1 (2016): 44–54.

36. S. F. Brosnan and F.B.M. de Waal, "Monkeys reject unequal pay," *Nature* 425, no. 6955 (2003): 297–299.

37. K. Jensen, J. Call, and M. Tomasello, "Chimpanzees are vengeful but not spiteful," *Proceedings of the National Academy of Sciences* 104, no. 32 (2007): 13046–13050.

38. M. D. Hauser, "Costs of deception: Cheaters are punished in Rhesus monkeys (Macaca mulatta)," *Proceedings of the National Academy of Sciences* 89, no. 24 (1992): 12137–12139.

39. I. Adachi, H. Kuwahata, and K. Fujita, "Dogs recall their owner's face upon hearing the owner's voice," *Animal Cognition* 10, no. 1 (2007): 17–21; S. Takagi et al., "Cats match voice and face: Cross-modal representation of humans in cats (Felis catus)," *Animal Cognition* 22, no. 5 (2019): 901–906.

40. J. Bräuer and D. Blasi, "Dogs display owner-specific expectations based on olfaction," *Scientific Reports* 11, no. 1 (2021): 3291.

41. L. Proops, K. McComb, and D. Reby, "Cross-modal individual recognition in domestic horses (Equus caballus)," *Proceedings of the National Academy of Sciences* 106, no. 3 (2009): 947–951.

42. J. N. Bruck, S. F. Walmsley, and V. M. Janik, "Cross-modal perception of identity by sound and taste in bottlenose dolphins," *Science Advances* 8, no. 20 (2022): eabm7684.

43. M. Loconsole, G. Stancher, and E. Versace, "Crossmodal association between visual and acoustic cues in a tortoise (Testudo hermanni)," *Biology Letters* 19, no. 7 (2023): 20230265.

44. C. Solvi, S. G. Al-Khudhairy, and L. Chittka, "Bumble bees display crossmodal object recognition between visual and tactile senses," *Science* 367, no. 6480 (2020): 910–912.

45. K. J. Park et al., "An unusual case of care-giving behavior in wild long-beaked common dolphins (Delphinus capensis) in the East Sea," *Marine Mammal Science* 29, no. 4 (2012): E508–E514.

46. S. E. Turner, L. Gould, and D. A. Duffus, "Maternal behavior and infant congenital limb malformation in a free-ranging group of *Macaca fuscata* on Awaji Island, Japan," *International Journal of Primatology* 26, no. 6 (2005): 1435–1457.

47. T. Matsumoto et al., "An observation of a severely disabled infant chimpanzee in the wild and her interactions with her mother," *Primates* 57, no. 1 (2016): 3–7.

48. C. Boesch and H. Boesch-Achermann, *The Chimpanzees of the Tai Forest: Behavioural Ecology and Evolution* (Oxford; New York: Oxford University Press, 2000); K. A. Cronin et al. "Behavioral response of a chimpanzee mother toward her dead infant," *American Journal of Primatology* 73, no. 5 (2011): 415–421; A. De Marco, R. Cozzolino, and B. Thierry, "Prolonged transport and cannibalism of mummified infant remains by a Tonkean Macaque mother," *Primates* 59, no. 1 (2018): 55–59; A. De Marco, R. Cozzolino, and B. Thierry, "Responses to a dead companion in a captive group of tufted capuchins (Sapajus apella)," *Primates* 61, no. 1 (2020): 111–117; P. J. Fashing et al., "Death among Geladas (Theropithecus gelada): A broader perspective on mummified infants and primate thanatology," *American Journal of Primatology* 73, no. 5 (2011): 405–409; J. D. Pruetz et al., "Intragroup lethal aggression in West African chimpanzees (Pan troglodytes verus): Inferred killing of a former alpha male at Fongoli, Senegal," *International Journal of Primatology* 38, no. 1 (2017): 31–57; B. Yang, J. R. Anderson, and B.-G. Li, "Tending a dying adult in a wild multilevel primate society," *Current Biology* 26, no. 10 (2016): R403–R404.

49. F. A. Stewart, A. K. Piel, and R. C. O'Malley, "Responses of chimpanzees to a recently dead community member at Gombe National Park, Tanzania," *American Journal of Primatology* 74, no. 1 (2012): 1–7.

50. Bearzi et al., "Whale and dolphin behavioural responses."

51. C. Trapanese et al., "Prolonged care and cannibalism of infant corpse by relatives in semi-free-ranging capuchin monkeys," *Primates* 61, no. 1 (2020): 41–47.

52. S. Z. Goldenberg and G. Wittemyer, "Elephant behavior toward the dead: A review and insights from field observations," *Primates* 61, no. 1 (2020): 119–128.

53. Brosnan and Vonk, "Nonhuman primate responses."

54. For a review, see T. R. Zentall, "Animals represent the past and the future," *Evolutionary Psychology* 11, no. 3 (2013): 573–590.

55. N. S. Clayton and A. Dickinson, "Episodic-like memory during cache recovery by scrub jays," *Nature* 395, no. 6699 (1998): 272–274.

56. C. R. Raby et al., "Planning for the future by western scrub-jays," *Nature* 445, no. 7130 (2007): 919–921.

57. S.P.C. Correia, A. Dickinson, and N. S. Clayton, "Western scrub-jays anticipate future needs independently of their current motivational state," *Current Biology* 17, no. 10 (2007): 856–861.

58. A. De Marco, R. Cozzolino, and B. Thierry, "Coping with mortality: Responses of monkeys and great apes to collapsed, inanimate and dead conspecifics," *Ethology Ecology & Evolution* 34, no. 1 (2022): 1–50.

59. Gonçalves and Carvalho, "Death among Primates"; Gonçalves and Biro, "Comparative Thanatology."

60. G. Teleki, "Group response to the accidental death of a chimpanzee in Gombe National Park, Tanzania," *Folia Primatologica* 20, nos. 2–3 (1973): 81–94.

61. M. Leroux et al., "First observation of a chimpanzee with albinism in the wild: Social interactions and subsequent infanticide," *American Journal of Primatology* 84, no. 6 (2022): e23305.

62. Brosnan and Vonk, "Nonhuman Primate Responses."

63. D. P. Watts, "Responses to dead and dying conspecifics and heterospecifics by wild mountain gorillas (Gorilla beringei beringei) and chimpanzees (Pan troglodytes schweinfurthii)," *Primates* 61, no. 1 (2020): 55–68.

64. Watts, "Responses to dead and dying conspecifics," pp. 60–61.

65. A. De Marco, R. Cozzolino, and B. Thierry, "Responses to a dead companion in a captive group of tufted capuchins (Sapajus apella)," *Primates* 61, no. 1 (2020): 111–117.

66. R. Moore, "There is a moral argument for keeping great apes in zoos," *Aeon* (2017). Available at: https://aeon.co/ideas/there-is-a-moral-argument-for-keeping-great-apes-in-zoos (accessed February 15, 2021).

67. Z. Goldsborough et al., "Do chimpanzees (Pan troglodytes) console a bereaved mother?," *Primates* 61, no. 1 (2020): 93–102.

68. C. Allen and M. D. Hauser, "Concept attribution in nonhuman animals: Theoretical and methodological problems in ascribing complex mental processes," *Philosophy of Science* 58, no. 2 (1991): 221–240.

69. K. N. Swift and J. M. Marzluff, "Wild American crows gather around their dead to learn about danger," *Animal Behaviour* 109 (2015): 187–197.

70. G. Ohashi and T. Matsuzawa, "Deactivation of snares by wild chimpanzees," *Primates* 52, no. 1 (2011): 1–5.

71. C. Crockford et al., "Wild chimpanzees inform ignorant group members of danger," *Current Biology* 22, no. 2 (2012): 142–146.

72. K. J. Hockings, J. R. Anderson, and T. Matsuzawa, "Road crossing in chimpanzees: A risky business," *Current Biology* 16, no. 17 (2006): R668–R670.

73. I. R. Clark et al., "A preliminary analysis of wound care and other-regarding behavior in wild chimpanzees at Ngogo, Kibale National Park, Uganda," *Primates* 62, no. 5 (2021): 697–702; Y. Sato, S. Hirata, and F. Kano, "Spontaneous attention and psycho-physiological responses to others' injury in chimpanzees," *Animal Cognition* 22, no. 5 (2019): 807–823.

74. H. F. Harlow, R. O. Dodsworth, and M. K. Harlow, "Total social isolation in monkeys," *Proceedings of the National Academy of Sciences of the United States of America* 54, no. 1 (1965): 90–97.

75. G. W. Kraemer and A. S. Clarke, "The behavioral neurobiology of self-injurious behavior in rhesus monkeys," *Progress in Neuro-Psychopharmacology and Biological Psychiatry* 14 (1990): S141–S168.

76. I. H. Jones and B. M. Barraclough, "Auto-mutilation in animals and its relevance to self-injury in man," *Acta Psychiatrica Scandinavica* 58, no. 1 (1978): 40–47.

77. S. M. McDonnell, "Practical review of self-mutilation in horses," *Animal Reproduction Science* 107, no. 3 (2008): 219–228.

78. U. A. Luescher, D. B. McKeown, and J. Halip, "Stereotypic or obsessive-compulsive disorders in dogs and cats," *Veterinary Clinics of North America: Small Animal Practice* 21, no. 2 (1991): 401–413.

79. D. Peña-Guzmán, "Can nonhuman animals commit suicide?," *Animal Sentience: An Interdisciplinary Journal on Animal Feeling* 2, no. 20 (2017). DOI: 10.51291/2377-7478.1201

80. C. Riley, "The dolphin who loved me: The NASA-funded project that went wrong," *The Guardian* (2014). Available at: http://www.theguardian.com/environment/2014/jun/08/the-dolphin-who-loved-me (accessed May 10, 2021).

81. B. J. King, *How Animals Grieve* (Chicago: University of Chicago Press, 2013), p. 265.

82. C. Rutz, M. Bronstein, A. Raskin, S. C. Vernes, K. Zacarian, and D. E. Blasi, "Using machine learning to decode animal communication," *Science* 381, no. 6654 (2023): 152–155.

Chapter 7: The Opossum Who Was Both Dead and Alive

1. A. L. Gardner, "Virginia opossum: Didelphis virginiana," in J. A. Chapman and G. A. Feldhamer (eds.), *Wild Mammals of North America* (Baltimore, MD: Johns Hopkins University Press, 1982), p. 6.

2. E. N. Francq, "Behavioral aspects of feigned death in the opossum Didelphis marsupialis," *American Midland Naturalist* 81, no. 2 (1969): 556–568; G. W. Gabrielsen and E. N. Smith, "Physiological responses associated with feigned death in the American opossum," *Acta Physiologica Scandinavica* 123, no. 4 (1985): 393–398.

3. The need to look toward violent contexts and interspecific relations to find a concept of death in nature was previously defended in S. Monsó and A. J. Osuna-Mascaró, "Death is common, so is understanding it: The concept of death in other species," *Synthese* 199, no. 1 (2021): 2251–2275.

4. J. M. Gómez et al., "The phylogenetic roots of human lethal violence," *Nature* 538, no. 7624 (2016): 233–237.

5. A. Gonçalves and S. Carvalho, "Death among primates: A critical review of non-human primate interactions towards their dead and dying," *Biological Reviews* 94, no. 4 (2019): 1502–1529.

6. F.B.M. de Waal, *Chimpanzee Politics: Power and Sex among Apes* (Baltimore, MD: Johns Hopkins University Press, 2007), p. 211.

7. S.S.K. Kaburu, S. Inoue, and N. E. Newton-Fisher, "Death of the alpha: Within-community lethal violence among chimpanzees of the Mahale Mountains National Park," *American Journal of Primatology* 75, no. 8 (2013): 789–797.

8. Kaburu et al., "Death of the alpha," p. 794.

9. T. Nishida, "The death of Ntologi, the unparalleled leader of M group," *Pan Africa News* 3, no. 1 (1996): 9–11.

10. J. Grinnell, C. Packer, and A. E. Pusey, "Cooperation in male lions: Kinship, reciprocity or mutualism?," *Animal Behaviour* 49, no. 1 (1995): 95–105.

11. D. Mech and L. Boitani, "Wolf social ecology," in D. Mech and L. Boitani (eds.), *Wolves: Behavior, Ecology, and Conservation* (Chicago: University of Chicago Press, 2003), pp. 1–34.

12. J. Gros-Louis, S. Perry, and J. H. Manson, "Violent coalitionary attacks and intraspecific killing in wild white-faced capuchin monkeys (Cebus capucinus)," *Primates* 44, no. 4 (2003): 341–346, reference at p. 343.

13. D. P. Watts et al., "Lethal intergroup aggression by chimpanzees in Kibale National Park, Uganda," *American Journal of Primatology* 68, no. 2 (2006): 161–180.

14. D. Peterson and R. Wrangham, *Demonic Males: Apes and the Origins of Human Violence* (Boston: Mariner Books, 1997), pp. 5–6.

15. Peterson and Wrangham, *Demonic Males*, p. 14.

16. Peterson and Wrangham, *Demonic Males*, p. 6.

17. A. K. Brown et al., "Infanticide by females is a leading source of juvenile mortality in a large social carnivore," *American Naturalist* 198, no. 5 (2021): 642–652.

18. A. E. Lowe et al., "Intra-community infanticide in wild, eastern chimpanzees: A 24-year review," *Primates* 61, no. 1 (2020): 69–82.

19. In what follows, I use the analyses by S. B. Hrdy, "Infanticide among animals: A review, classification, and examination of the implications for the reproductive

strategies of females," *Ethology and Sociobiology* 1, no. 1 (1979): 13–40; C. van Schaik, "Infanticide by male primates: The sexual selection hypothesis revisited," in C. van Schaik and C. H. Janson (eds.), *Infanticide by Males and Its Implications* (Cambridge: Cambridge University Press, 2000), pp. 27–60; M. F. Li, "Infanticide," in J. Vonk and T. Shackelford (eds.), *Encyclopedia of Animal Cognition and Behavior* (Cham: Springer International, 2019). DOI: 10.1007/978-3-319-47829-6 _573-2.

20. E. R. Vogel and A. Fuentes-Jiménez, "Rescue behavior in white-faced capuchin monkeys during an intergroup attack: Support for the Infanticide Avoidance Hypothesis," *American Journal of Primatology* 68, no. 10 (2006): 1012–1016.

21. Van Schaik, "Infanticide by male primates."

22. Hrdy, "Infanticide among animals."

23. D. Lukas and E. Huchard, "The evolution of infanticide by females in mammals," *Philosophical Transactions of the Royal Society B: Biological Sciences* 374, no. 1780 (2019): 20180075.

24. Hrdy, "Infanticide among animals."

25. H. M. Bruce, "An exteroceptive block to pregnancy in the mouse," *Nature* 184, no. 4680 (1959): 105–105.

26. G. Perrigo, L. Belvin, and F. S. Vom Saal, "Time and sex in the male mouse: Temporal regulation of infanticide and parental behavior," *Chronobiology International* 9, no. 6 (1992): 421–433.

27. R. W. Elwood and D. S. Stolzenberg, "Flipping the parental switch: From killing to caring in male mammals," *Animal Behaviour* 165 (2020): 133–142.

28. C. Rudolf von Rohr et al., "'Chimpanzees' bystander reactions to infanticide," *Human Nature* 26, no. 2 (2015): 143–160.

29. F.B.M. de Waal, "Natural normativity: The 'is' and 'ought' of animal behavior," *Behaviour* 151, nos. 2–3 (2014): 185–204, reference on page 189.

30. Van Schaik, "Infanticide by male primates."

31. Hrdy, "Infanticide among animals."

32. Van Schaik, "Infanticide by male primates," p. 52.

33. J. R. Towers et al., "Infanticide in a mammal-eating killer whale population," *Scientific Reports* 8, no. 1 (2018): 1–8.

34. L.J.N. Brent et al., "Ecological knowledge, leadership, and the evolution of menopause in killer whales," *Current Biology* 25, no. 6 (2015): 746–750.

35. Towers et al., "Infanticide," p. 6.

36. Lukas and Huchard, "The evolution of infanticide."

37. Brown et al., "Infanticide by females."

38. Hrdy, "Infanticide among animals," p. 16.

39. J. Böhm et al., "The Venus flytrap *Dionaea muscipula* counts prey-induced action potentials to induce sodium uptake," *Current Biology* 26, no. 3 (2016): 286–295.

40. P. Calvo, "Plantae," in J. Vonk and T. Shackelford (eds.), *Encyclopedia of Animal Cognition and Behavior* (Cham: Springer International, 2018). DOI: 10.1007/978-3-319-47829-6_1812-1. For an excellent introduction to the plant cognition debate, see also: P. Calvo and N. Lawrence, *Planta Sapiens: Unmasking Plant Intelligence* (London: The Bridge Street Press, 2022).

41. M. Biben, "Predation and predatory play behaviour of domestic cats," *Animal Behaviour* 27 (1979): pp. 81–94, reference on p. 86.

42. Peterson and Wrangham, *Demonic Males*, p. 216.

43. S. H. Ferguson, J. W. Higdon, and K. H. Westdal, "Prey items and predation behavior of killer whales (Orcinus orca) in Nunavut, Canada based on Inuit hunter interviews," *Aquatic Biosystems* 8, no. 3 (2012), see pp. 7, 11.

44. M. P. Cotter, D. Maldini, and T. A. Jefferson, "'Porpicide' in California: Killing of harbor porpoises (Phocoena phocoena) by coastal bottlenose dolphins (Tursiops truncatus)," *Marine Mammal Science* 28, no. 1 (2012): E1–E15.

45. Cotter et al., "'Porpicide,'" p. E7.

46. G. B. Schaller, *The Serengeti Lion: A Study of Predator-Prey Relations* (Chicago: University of Chicago Press, 1976), p. 389.

47. S. A. Temple, "Do predators always capture substandard individuals disproportionately from prey populations?," *Ecology* 68, no. 3 (1987): 669–674.

48. Temple, "Do predators."

49. Biben, "Predation."

50. Schaller, *The Serengeti Lion*.

51. Schaller, *The Serengeti Lion*.

52. Biben, "Predation," p. 86.

53. L. Crisler "Observations of wolves hunting caribou," *Journal of Mammalogy* 37, no. 3 (1956): 337–346.

54. A. V. Shubkina, A. S. Severtsov, and K. V. Chepeleva, "Factors influencing the hunting success of the predator: A model with sighthounds," *Biology Bulletin* 39, no. 1 (2012): 65–76.

55. C. E. Krumm et al., "Mountain lions prey selectively on prion-infected mule deer," *Biology Letters* 6, no. 2 (2010): 209–211.

56. Schaller, *The Serengeti Lion*, p. 241.

57. M. Á. Gómez-Serrano, "Broken wing display," in J. Vonk and T. Shackelford (eds.), *Encyclopedia of Animal Cognition and Behavior* (Cham: Springer International, 2018). DOI: 10.1007/978-3-319-47829-6_2007-2.

58. M. Á. Gómez-Serrano and P. López-López, "Deceiving predators: Linking distraction behavior with nest survival in a ground-nesting bird," *Behavioral Ecology* 28, no. 1 (2017): 260–269.

59. C. Ristau, "Aspects of the cognitive ethology of an injury-feigning bird, the piping plover," in C. Ristau (ed.), *Cognitive Ethology: The Minds of Other Animals* (Hove; Hillsdale, NJ: Lawrence Erlbaum Associates, 1991), pp. 91–126.

60. Schaller, *The Serengeti Lion*, p. 387.

61. F. B. Bercovitch, "A comparative perspective on the evolution of mammalian reactions to dead conspecifics," *Primates* 6, no. 1 (2020): 21–28.

62. Schaller, *The Serengeti Lion*, p. 388.

63. This idea was first defended in Monsó and Osuna-Mascaró, "Death is common."

64. J. Kalaf et al., "Sexual trauma is more strongly associated with tonic immobility than other types of trauma—A population based study," *Journal of Affective Disorders* 215 (2017): 71–76.

65. H. Nishino, "Motor output characterizing thanatosis in the cricket Gryllus bimaculatus," *Journal of Experimental Biology* 207, no. 22 (2004): 3899–3915.

66. H. H. Vogel Jr., "Observations on social behavior in turkey vultures," *Auk* 67, no. 2 (1950): 210–216.

67. P. T. Gregory, L. A. Isaac, and R. A. Griffiths, "Death feigning by grass snakes (*Natrix natrix*) in response to handling by human 'predators,'" *Journal of Comparative Psychology* 12, no. 2 (2007): 123–129.

68. G. M. Burghardt, "Cognitive ethology and critical anthropomorphism: A snake with two heads and hognose snakes that play dead," in *Cognitive Ethology: The Minds of Other Animals* (Hove; Hillsdale, NJ: Lawrence Erlbaum Associates, 1991), pp. 53–90.

69. Here I'm following the analysis by B. Rojas and E. Burdfield-Steel, "Predator defense," in J. Vonk and T. Shackelford (eds.), *Encyclopedia of Animal Cognition and Behavior* (Cham: Springer International, 2017). DOI: 10.1007/978-3-319-47829-6_708-1.

70. Toledo et al., "Is it all death feigning?"; T. Miyatake et al., "Tonically immobilized selfish prey can survive by sacrificing others," *Proceedings of the Royal Society B: Biological Sciences* 276, no. 1668 (2009): 2763–2767.

71. S. M. Rogers and S. J. Simpson, "Thanatosis," *Current Biology* 24, no. 21 (2014): R1031–R1033.

72. D. L. Cassill, K. Vo, and B. Becker, "Young fire ant workers feign death and survive aggressive neighbors," *Naturwissenschaften* 95, no. 7 (2008): 617–624.

73. Toledo et al., "Is it all death feigning?."

74. A. Honma, S. Oku, and T. Nishida, "Adaptive significance of death feigning posture as a specialized inducible defence against gape-limited predators," *Proceedings of the Royal Society B: Biological Sciences* 273, no. 1594 (2006): 1631–1636.

75. R. K. Humphreys and G. D. Ruxton, "A review of thanatosis (death feigning) as an anti-predator behaviour," *Behavioral Ecology and Sociobiology* 72, no. 22 (2018). DOI: 10.1007/s00265-017-2436-8.

76. Gardner, "Virginia opossum."

77. J. R. Anderson, H. Yeow, and S. Hirata, "Putrescine—a chemical cue of death—is aversive to chimpanzees," *Behavioural Processes* 193 (2021): 104538.

78. Rojas and Burdfield-Steel, "Predator defense"; Honma et al., "Adaptive significance."
79. Honma et al., "Adaptive significance."
80. Gregory et al., "Death feigning by grass snakes."
81. Humphreys and Ruxton, "A review of thanatosis."
82. J. Skelhorn, "Avoiding death by feigning death," *Current Biology* 28, no. 19 (2018): R1135–R1136.
83. Gregory et al., "Death feigning by grass snakes"; Burghardt, "Cognitive ethology"; Humphreys and Ruxton, "A review of thanatosis."
84. Humphreys and Ruxton, "A review of thanatosis."

Image Credits

1. © Monica Szczupider.
2. These images originally appeared on pages 917 and 919 of E.J.C. van Leeuwen et al., "Chimpanzees' responses to the dead body of a 9-year-old group member," *American Journal of Primatology* 78, no. 9 (2016): 914–922.
3. This image originally appeared on page 6 of D. J. Povinelli and J. Vonk, "We don't need a microscope to explore the chimpanzee's mind," *Mind and Language* 19, no. 1 (2004): 1–28.
4. These are screenshots of videos 2 and 3 in the supplementary material of J. D. Negrey and K. E. Langergraber, "Corpse-directed play parenting by a sterile adult female chimpanzee," *Primates* 61, no. 1 (2020): 29–34.
5. Nigel Dowsett / Alamy Stock Photo.
6. This image originally appeared in A. Porter et al. (2018). "Mountain gorilla juvenile male Segasira with his mother Tuck's corpse." https://doi.org/10.6084/m9.figshare.6198587.
7. This image originally appeared on page 96 in I. Douglas-Hamilton et al., "Behavioural reactions of elephants towards a dying and deceased matriarch," *Applied Animal Behaviour Science* 100, nos. 1–2 (2006): 87–102.
8. These images are photographed by Arianna de Marco and originally appeared on page 57 of A. de Marco, R. Cozzolino, and B. Thierry, "Prolonged transport and cannibalism of mummified infant remains by a Tonkean macaque mother," *Primates* 59, no. 1 (2018): 55–59.
9. These images are photographed by Dora Biro and originally appeared on page 246 of D. Biro, "Chimpanzee mothers carry the mummified remains of their dead infants: Three case reports

from Bossou," in T. Matsuzawa, T. Humle, and Y. Sugiyama (eds.), *The Chimpanzees of Bossou and Nimba* (Springer Japan, Primatology Monographs, 2011), pp. 241–250.

10. These illustrations originally appeared on page 1313 of G. Vallortigara, L. Regolin, and F. Marconato, "Visually inexperienced chicks exhibit spontaneous preference for biological motion patterns," *PLOS Biology* 3, no. 7 (2005): e208.

11. This image and illustrations originally appeared on page E510 of K. J. Park et al., "An unusual case of care-giving behavior in wild long-beaked common dolphins (Delphinus capensis) in the East Sea," *Marine Mammal Science* 29, no. 4 (2012): E508–E514.

12. This image is photographed by Maël Leroux and originally appeared on page 4 of M. Leroux et al., "First observation of a chimpanzee with albinism in the wild: Social interactions and subsequent infanticide," *American Journal of Primatology* (2021): e23305. DOI: 10.1002/ajp.23305.

13. © Sönke Scherzer.

14. These images are photographed by Miguel Ángel Gómez-Serrano and originally appeared on page 2 of M. A. Gómez-Serrano, "Broken wing display," in J. Vonk and T. Shackelford (eds.), *Encyclopedia of Animal Cognition and Behavior* (Cham: Springer International, 2018), pp. 1–3. DOI: 10.1007/978-3-319-47829-6_2007-2.

15. This image originally appeared on page 1984 of L. F. Toledo, I. Sazima, and C.F.B. Haddad, "Is it all death feigning? Case in anurans," *Journal of Natural History* 44, nos. 31–32 (2010): 1979–1988.

16. These images originally appeared on page 125 of P. T. Gregory, L. A. Isaac, and R. A. Griffiths, "Death feigning by grass snakes (Natrix natrix) in response to handling by human 'predators,'" *Journal of Comparative Psychology* 121, no. 2 (2007): 123–129.

17. This image originally appeared on page 1984 of L. F. Toledo, I. Sazima, and C.F.B. Haddad, "Is it all death feigning? Case in anurans," *Journal of Natural History* 44, nos. 31–32 (2010): 1979–1988.

18. These images originally appeared on page 1633 of A. Honma, S. Oku, and T. Nishida, "Adaptive significance of death feigning posture as a specialized inducible defence against gape-limited predators," *Proceedings of the Royal Society B: Biological Sciences* 273, no. 1594 (2006): 1631–1636.

Index

Page numbers in italics indicate illustrations.

affiliative behaviors, 83–84, 140–41. *See also* grief
Allen, Colin, 43
Anderson, James, 80, 86, 199
Andrews, Kristin, 46
anecdotal method, 39–42, 44. *See also* animal minds
animal anecdotes. *See* Dorothy (chimpanzee); Evalyne (Tonkean macaque); Flint (chimpanzee); Flipper (dolphin); Godi (chimpanzee); Grace (elephant); Lucy (chimpanzee); Luit (chimpanzee); Moni (chimpanzee); Pablo (monkey); Pansy (chimpanzee); Peter (dolphin); Pimu (chimpanzee); Segasira (gorilla); Simba (gorilla); Tahlequah (orca); Willa (cat); ZBD (monkey)
animal minds: beliefs and, 13–17; existence of, 17–18; methods to study, 33–42. *See also* belief; concept of death; concepts
animal psychology, 6
animal self-awareness. *See* mirror self-recognition test
anomalous behavior, 127–30. *See also* entities

anthropectomy, 45–47, 51–52
anthropocentrism: bias and, 5–6, 45–48, 50–52, 209; emotional, 51, 79–84, 91, 102–5, 116, 149, 153, 188; intellectual, 51, 56–63, 75, 82; mortality and, 207–8. *See also* comparative thanatology; grief
anthropomorphism: bias and, 45–47, 50–52; comparative thanatology and, 32–33, 44. *See also* comparative thanatology
anti-consumption mechanisms, 195–96. *See also* defense mechanism; predation; thanatosis
anti-detection mechanisms, 193. *See also* defense mechanism; predation; thanatosis
anti-recognition mechanisms, 193–94, 198–200, 202. *See also* defense mechanism; predation; thanatosis
anti-subjugation mechanisms, 194–95, 200–202. *See also* defense mechanism; predation; thanatosis
ants: concept of death and, 11, 13, 23–25; defense mechanisms, 195; necrophoresis and, 11–12, 23–25. *See also* concept of death

INDEX

apes, 125, 135, 140. *See also* bonobos; chimpanzees; gorillas; primates
aposematism, 194. *See also* defense mechanism
Appleby, Rob, 92
association, 143. *See also* minimal concept of death

Baker, Lucy, 107
Bates, Lucy, 39–41
Batesian mimicry, 194, 204. *See also* defense mechanism
Bearzi, Giovanni, 113, 131
Belshaw, Christopher, 59
beluga whale, 39–40
Bercovitch, Fred, 113, 116, 188
bias. *See* anthropocentrism; anthropomorphism; comparative thanatology
Biben, Maxeen, 177, 183
biological movement, 121–22, *123*. *See also* entities
birds, 185–86; crows, 144; scrub jays, 133–34
Biro, Dora, 43
Blake, Randolph, 121–22
bonobos, 125–26. *See also* apes; primates
Bradley, Ben, 59
Bronte (chimpanzee). *See* Lucy (chimpanzee)
Brosnan, Sarah, 137–38
Bruce, Hilda M. *See* Bruce effect
Bruce effect, 168. *See also* infanticide
bumblebees, 126
Burghardt, Gordon, 191
Burkart, Judith, 169
Byrne, Richard, 39–41

Call, Josep, 125
cannibalism, 95, 102–3

capuchin monkeys, 126, 131, 140, 165. *See also* monkeys; primates
carcass. *See* corpses
carrying techniques, 99. *See also* corpse carrying
Carvalho, Susana, 57–58
cats, 91, 121–22, 126, 147, 152, 183. *See also* animal anecdotes
cetaceans, 92, 131, 179–80, 183. *See also* beluga whale; dolphins; orcas
Challenger, Melanie, 206
Cheney, Dorothy, 38
chicks, 122
chimpanzees: coalitional attacks and, 158–60, 162–63; compensatory care of, 129; corpses and, 131, 136–37, *138*, 138–39; danger and, 144–45; expectations within, 126; of Gombe, 85; grief and, 1, 2, 80–81, 140–41; hunting and, 178–79; of Kibale National park, 53–56; reactions to death and, 28–30, 29, 199; self-awareness and, 49; theory of mind and, 34–37, *35*. *See also* animal anecdotes; apes; primates
Cigman, Ruth, 58, 60
Clayton, Nicola, 133
cleaner wrasse (fish), 50–51
coalitional attacks, 158–63. *See also* violence
cognition: irreversibility and, 132–39; non-functionality and, 121. *See also* concept of death; minimal concept of death; predation
cognitive mechanisms, 3
communication, 33, 110–11, 149. *See also* experience; language
comparative psychology, 3, 17, 45–46, 149. *See also* comparative thanatology
comparative thanatology: bias and, 5–6, 32–33, 44–48, 51–52, 56, 81–82,

137–38; definition of, 2–3; evolution and, 48; human values and, 48, 84; studies in, 42–44, 79. *See also* affiliative behaviors; anthropocentrism; anthropomorphism; interspecific violence
concept of death: animals and, 59–60, 62–63, 189, 207–10; categorial desires and, 58; cognitive reactions and, 22, 24, 25–26, 28–30, 75, 96–97, 109–10, 143–44, 155–57, 169, 171–78, 183, 186, 203, 209; example of, 8–10; grief and, 90–91, 103–5; harm and, 58–59; holy trinity of, 109–17, 150, 178; humans and, 61–62, 207–10; overintellectualization of, 57, 60–61, 82; stereotypical reactions and, 22–23, 24, 26–28, 64, 75, 116, 166–67, 199, 209; subcomponents of, 64; theory of mind and, 67–68. *See also* cognition; emotion; experience; minimal concept of death; predation; sociality; thanatosis
concepts: animals with, 19–20, 22; behavioral responses and, 21–22; conditions of, 63–64; semantic content of, 20–21; sensory stimulus and, 21. *See also* animal minds; minimal concept of death
conceptual characterizations, 56
corpse carrying, 55–56, 91–103, 93. *See also* mother-infant bond; primates
corpses, 130–36. *See also* chimpanzees; concept of death; corpse carrying; elephants; necrophilia; necrophobia; necrophoresis; primates; thanatosis
Cotter, Mark, 180
Cozzolino, Roberto, 92, 135
Crisler, Lois, 184

cross-modal associations, 126
crows, 144

Davidson, Donald, 13–17, 216n9. *See also* belief
defense mechanism, 151–52, 190, 193–95. *See also* thanatosis
de Kort, Dante, 89
de Marco, Arianna, 57, 92, 135, 140
de Unamuno, Miguel, 61–62
de Waal, Frans, 159, 169
dingos, 92. *See also* dogs
dogs, 50, 77–79, 126, 147, 184–85. *See also* dingos; wolves
dolphins, 98, 126–27, 128, 148, 180. *See also* animal anecdotes; cetaceans
Dorothy (chimpanzee), 1, 2. *See also* chimpanzees
Douglas-Hamilton, Iain, 87
Duffus, David, 128

ecological validity, 35–38. *See also* experimental method; observational method
Eddy, Timothy, 34–37. *See also* chimpanzees
elephants: concept of death within, 150; empathy within, 41–42, 106; grief and, 87, 88; relationship with death, 106–9, 112–14; smell and, 131–32; studies of, 107–8. *See also* animal anecdotes
emotion, 111, 114–15, 117–18, 139, 150. *See also* concept of death; grief; predation; sociality
entities: animate vs inanimate, 121–22; expectations toward, 122, 124–29, 134–36. *See also* anomalous behavior; biological movement
ethics, 10, 58. *See also* morality

ethology, 3
Evalyne (Tonkean macaque), 92–95, 93. *See also* primates
evolutionary biology, 5
experience, 110–11, 115, 118–20, 139, 150, 187–89. *See also* communication; concept of death; non-functionality; predation; sociality
experimental method, 33–37, 43. *See also* animal minds

Ferguson, Steven, 179
fission-fusion dynamics, 113–14, 162. *See also* sociality
Flint (chimpanzee), 85. *See also* chimpanzees
Flipper (dolphin), 148. *See also* dolphins
frogs, 191, *192*, *196–97*
Fuentes-Jiménez, Alexander, 165
funerary practice, 144

Gallup, Gordon, 49–50. *See also* chimpanzees
generalized aggression, 165–66. *See also* infanticide
gerbils. *See* rodents
Gillies, Alasdair, 80
giraffes, 89
Godi (chimpanzee), 162. *See also* chimpanzees
Goldsborough, Zoë, 140
Gonçalves, André, 43, 57–58
Goodall, Jane, 37, 85. *See also* chimpanzees
gorillas, 85–86, 102–3. *See also* animal anecdotes; apes; primates
Gould, Lisa, 128
Grace (elephant), 87, *88*, 107, 127. *See also* elephants
Gregory, Patrick, 200

grief, 84–91, 103–5, 117, 140–41, 208. *See also* affiliative behaviors; anthropocentrism; chimpanzees; concept of death; elephants; emotion; Tahlequah (orca)
Gros-Lous, Julie, 161

Harlow, Harry, 147
Harman, Elizabeth, 59
Hauser, Marc, 43
Heidegger, Martin, 60
Higdon, Jeff, 179
Hirata, Satoshi, 199
horses, 147
How to Be Animal (Challenger), 206
Hrdy, Sarah Blaffer, 173
Huchard, Elise, 172
Humphreys, Rosalind, 198, 200
hunting. *See* predation
Huss, Brian, 46
hyenas, 173

inductive generalization, 143. *See also* minimal concept of death
infanticide, 163–73. *See also* generalized aggression; proximate causes; sexual selection hypothesis; violence
Inoue, Sara, 159–60
interspecific violence, 153. *See also* violence
irreversibility, 132–33, 135–36. *See also* minimal concept of death; non-functionality

Japanese macaques, 43, 128–29. *See also* primates
Jefferson, Thomas, 180
Jones, Darryl, 92

Kaburu, Stefano, 159–60
Kahlenberg, Sonya, 54
kangaroos, 166
Kano, Fuhimiro, 125
Kentish plover, 185, 186. *See also* birds
King, Barbara, 90–91
Krumm, Caroline, 185
K-strategists, 101–3, 112–13, 115, 119

Langergraber, Kevin, 53
language, 16–17, 33, 60. *See also* communication
Leroux, Maël, 137
Li, Bao Guo, 86
Lilly, John, 148
lions, 160, 183, 185
Lock, Louise, 80
Lucy (chimpanzee), 53–56, 54, 76. *See also* chimpanzees
Luit (chimpanzee), 159. *See also* chimpanzees
Lukas, Dieter, 172

Maldini, Daniel, 180
Manson, Joseph, 161
Marcel (monkey), 161. *See also* capuchin monkeys
Marino, Lori, 40
Marler, Peter, 38
Marzluff, John, 144
Matsumoto, Takuya, 129
Matsuzawa, Tetsuro, 95–96, 98
McComb, Karen, 107
mimicry, 125, 194. *See also* defense mechanism
minimal concept of death: causality and, 71, 142–45, 150; cognitive reactions and, 75; as a concept, 63–64; concept of life and, 73–74; definition of, 72, 141; inevitability and, 70; inferences and, 74–75, 129–30, 184; irreversibility and, 69–71, 73–74, 76, 145, 201; non-functionality and, 65–69, 71, 73, 76, 135, 145, 201; personal mortality and, 70; reliability and, 72–73; sensory stimulus and, 75; universality and, 65, 70, 142–45, 150; unpredictability and, 71; variations and, 74. *See also* cognition; concept of death; concepts; irreversibility; non-functionality
mirror self-recognition test, 49–51
Moni (chimpanzee), 140–41. *See also* chimpanzees
monkeys, 38, 43, 86–87, 135, 140, 161–62. *See also* animal anecdotes; primates
Monsó, Susana, 216n6, 219n22. *See also* anthropocentrism; concept of death
morality, 5, 11. *See also* ethics
Moss, Cynthia, 107
mother-infant bond, 97–103. *See also* corpse carrying
Muller, Zoe, 89
Müllerian mimicry, 194. *See also* aposematism
multimodality, 130, 136

National Geographic, 1–2
natural concept of death, 142–45. *See also* minimal concept of death
natural selection, 18, 25, 129, 149, 164, 167, 202
necrophilia, 108, 131. *See also* corpses
necrophobia, 24, 27, 199. *See also* concept of death; corpses; necrophoresis; putrefaction; thanatosis
necrophoresis, 11–12, 22–24. *See also* concept of death; corpses; necrophobia; putrefaction
Negrey, Jacob, 53

Newton-Fisher, Nicholas, 159–60
Nishida, Toshishada, 160
nonbiological movement. *See* biological movement
non-functionality, 130–32, 134–35, 140, 185–86. *See also* experience; irreversibility; minimal concept of death; predation
notion of causality, 142–43, 149. *See also* minimal concept of death
notion of life, 142. *See also* minimal concept of death

O'Barry, Ric, 148
observational method, 37–38, 43. *See also* animal minds
O'Malley, Robert, 131
opossum, 151–53, 190–91, 199, 201, 204
orcas, 31–32, 171–72, 179–80. *See also* animal anecdotes; cetaceans
Osuna-Mascaró, Antonio J., 219n22. *See also* anthropocentrism

Pablo (monkey), 161. *See also* monkeys
Pansy (chimpanzee), 80–81. *See also* chimpanzees
Park, Kyum, 127
peccaries, 89–90
Perry, Susan, 161
personal mortality, 146–49. *See also* minimal concept of death
Peter (dolphin), 148. *See also* dolphins
Peterson, Dale, 162–63, 178
philosophy, 4
Piel, Alexander, 131
Pimu (chimpanzee), 160. *See also* chimpanzees
play-parenting, 54–56
Porter, Amy, 85

Povinelli, Daniel, 34–37. *See also* chimpanzees
predation: cognition and, 182–86; defense mechanisms against, 193–95; emotions and, 176–82; experience and, 187–89; incentives within, 181–82; intentionality within, 173–78, 183; stereotypical reactions within, 174–76. *See also* concept of death; prey; violence
predators. *See* predation
prey, 183–89. *See also* predation
primates: concept of death in, 57–58, 140; corpse carrying by, 55, 91–103, 100; humans and, 82, 83; infanticide within, 168–71; studies of, 2, 41, 82, 130–31, 135. *See also* apes; bonobos; capuchin monkeys; chimpanzees; gorillas; Japanese macaques; monkeys; Rhesus macaques
probability, 135
proximate causes, 164, 168, 173
putrefaction, 12–13, 136. *See also* concept of death; necrophobia; necrophoresis

rats, 26–27, 216n14
reactions. *See* concept of death
reasoning, 134. *See also* irreversibility
recategorization, 134. *See also* reasoning
Regan, Tom, 59
Reggente, Melissa, 98
resource competition hypothesis, 167, 172
Rhesus macaques, 126, 147. *See also* primates
rodents, 167–68
Rollin, Bernard, 58, 60
r-strategists, 101, 119–20

Rudolf von Rohr, Claudia, 169
Ruxton, Graeme, 198, 200

Schaller, George, 185, 187, 189
Schrödinger, Erwin, 152
scrub jays, 133–34
Segasira (gorilla), 85, 86, 102–3. See also gorillas
selection pressure, 203–5. See also thanatosis
self-mutilation, 147–48. See also personal mortality
sensory stimulus, 21–25, 64, 75, 174, 176, 199, 209. See also concept of death
sexual selection hypothesis, 167. See also infanticide
Seyfarth, Robert, 38
Simba (gorilla), 138–39. See also gorillas
Smith, Bradley, 92
snakes, 191, 192, 193
sociality, 113–16. See also concept of death; emotion; experience
social norms, 168–69
Stewart, Fiona, 131
stress, 166. See also infanticide
Sugiyama, Yukimaru, 43
suicidal behaviors, 147–49
survival, 124, 129
Swift, Kaeli, 144
Szczupider, Monica, 1, 2

tactile investigation, 131
Tahlequah (orca), 31–32, 44, 46–47, 55, 92, 138. See also orcas
thanatosis, 192; animal examples of, 191–93, 192; as anti-recognition mechanism, 198–200, 202; as anti-subjugation mechanism, 200, 202; concept of death and, 201–5; definition of, 190; evolution of, 202–5;

tonic immobility vs, 190–91, 193, 195, 196–97, 197–98, 201. See also defense mechanism
Thierry, Bernard, 92, 135
threat, 27–28, 40, 50, 98, 120, 124
Todorović, Ana, 99–101
tonic immobility. See thanatosis
tortoises, 126
Towers, Jared, 171
Trapanese, Cinzia, 131
turkey vultures, 191
Turner, Sarah, 128
typical behavior. See entities

ultimate causes, 164, 173

van Schaik, Carel, 166, 169, 171
Venus flytrap, 175
violence: intentionality within, 155–58, 157, 163, 165, 167, 170–73; intraspecific, 154–55; self-protection and, 189. See also interspecific violence; predation
Vogel, Erin, 165
Vogel, Howard, 191
Vonk, Jennifer, 137–38

Watson, Claire, 95–96, 98
Watts, David, 138
Westdal, Kristin, 179
Western scrub jays, 133–34
Whiten, Andrew, 41
Willa (cat), 91. See also cats
wolves, 78, 160, 184. See also dingos; dogs
Wrangham, Richard, 54, 162–63, 178

Yang, Bin, 86
Yeow, Hanling, 199

ZBD (monkey), 86–87